GEOFFREY MARNELL

# Essays on Technical Writing

Burdock Books

First published in Australia in 2016 by
Burdock Books
www.burdock.com.au
info@burdock.com.au

National Library of Australia Cataloguing-in-Publication entry

Creator:     Marnell, Geoffrey R., author.

Title:       Essays on Technical Writing / Geoffrey Marnell.

ISBN:        9780994366672 (paperback)

Notes:       Includes bibliographical references and index.

Subjects:    Technical writing.

             Communication of technical information.

Dewey Number: 808.0666

For *Melinda* ...

... with thanks to *Louise Correcha* for
editorial assistance

## About the author

Geoffrey Marnell has a masters degree and doctorate from the University of Melbourne, gained by research in philosophy at the universities of Melbourne and Oxford. He has published widely — on such topics as language, writing, psychology and mathematics.

Geoffrey tutored in philosophy at the University of Melbourne in the late 1970s and early 1980s before leaving academia to establish Abelard Consulting, a company that has, for close to 30 years, provided writing services, resources and training to organisations worldwide.

Geoffrey returned to the University in 2005 when, at the invitation of the English Department, he designed a course on technical writing and editing. He taught the course for nine years as part of the University's Publishing and Communications Program in the School of Culture and Communication.

In addition to language, Geoffrey's interests include literature, music, film and travel.

## By the same author

*Correct English: Reality or Myth?*
*Mindstretchers*
*Think About It!*
*Numberchains*

# Contents

# 1: Technical writing: what's in a name?

What is technical writing? How *technical* does technical writing need to be? And is the title of the profession suited to what we do?

Let's start off with a spot of surface analysis: at the simplest level, there are two broad views on what technical writing is: a prescriptivist view and a descriptivist view.

## Prescriptivist view

The prescriptivist view is that technical writing is writing about *technical* matters:

- "A working definition of technical ... communication should recognize the technical nature of the subject ..." (Zimmerman & Clark 19987, p. 3)
- "Technical writing is a form of written communication that conveys scientific and technical information in a clearly defined and accurate form." (Haydon 1995, p. 2)

A common definition of *technical* is "relating to or connected with the mechanical or industrial arts and the applied sciences" (*Macquarie Dictionary*). This is a definition that chimes well with many who are unaware of what our profession does. And yet it doesn't marry with the work that many technical writers actually do. It is stretching the meaning of *technical* to consider the following domains especially technical:

First published in *Words*, vol. 1, iss. 3, 2009. Material on the history of technical writing was adapted from a review of Tebeaux (1997) published in *Keyword* 1997.

- bookkeeping
- human resources procedures
- gaming.

And yet it is *technical* writers who are called on to write procedures explaining how to reconcile bank accounts, how to apply for long-service leave and how to play Tetris.

## Descriptivist view

The descriptivist view ignores the denotation and connotation of the word *technical* and looks instead at what actually goes on in our profession. By taking such a view, we find that:

- technical writers mostly engage in *procedural* writing (that is, *instructional* or *how-to* writing)
- the subject matter is sometimes *but not always* technical (as the word is commonly understood).

On the descriptivist view, it is the *type of writing* we do, not the subjects we write about, that identifies us as technical writers. Here are some definitions in that vein:

> "The purpose of technical communication is generally to instruct the reader (as opposed to scientific communication or journalism, which inform the reader). For example, online help teaches the reader how to perform various tasks using a software package; a car manual teaches the reader how to maintain and repair a car; and a set of illustrations teaches airline passengers how to behave in the event of an emergency."[1]

> "the large body of writing which may be called technical writing—how-to books or procedure manuals on a variety of topics: farming, gardening, animal husbandry, surveying, navigation, military science, accounting, recreation, estate management, household management, cooking, medicine, bee-keeping ..." (Tebeaux 1997, p. 93)

---

1.  Australian Society for Technical Communication website, http://astcvic.org.au/technical_communication/about.html. Viewed 15 January 2008.

Thus technical writing could be considered as the dissemination of practical knowledge (technical or otherwise), that is, knowledge about *how to do things*.

The descriptivist view is, I suggest, the better approach to defining our profession. Just as prescriptivist grammars risk irrelevancy by insisting on rules that few follow, a prescriptivist view of technical writing, with its definitional straight-jacket, risks irrelevancy. A parallel with science might be instructive. Science was once called *natural philosophy*, but to insist that what natural philosophy is should always be tied to the denotation and connotation of the word *philosophy* would have been futile. It would have led, as eventually happened, to a new name for the discipline.

Just as *natural philosophy* connotes a limited approach to knowledge — an *a priori* approach based on reason alone, unlike science with its *a posteriori* approach based on observation and experiment — *technical* writing connotes a limited approach to practical knowledge: limited to technical subjects. The former limitation, once fully recognised, led to a change of name: the advent of the term *science*. Perhaps, then, it is time for a new name for our profession, one that recognises the limitation of the term *technical*.

## The long history of technical writing

If we understand technical writing as the dissemination of practical knowledge (technical or otherwise), then it has a very long history. Elizabeth Tebeaux's *The Emergence of a Tradition: Technical Writing in the English Renaissance, 1475–1640* shows that technical writing came of age in the Renaissance. She provides excerpts from books on natural medicine, agriculture, navigation, surgical equipment, military combat and numerous other subjects, each immediately recognisable as technical writing. Figures 1.1 and 1.2 are reproduced from Tebeaux's book.

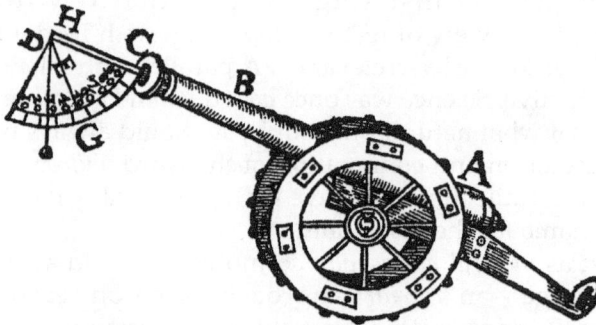

Firſte take the Quadrante, and put the rule of the Quadrante B into the mouth of the peece C and then putting
vp or downe the tayle of the peece A, till the plummet
G fall vpon the corner of the Quadrant at D, then looke
whatſoeuer you ſee right with the vpper ſtde of the Quadrante H, ſhall be leuell with the mouth of the peece,
and that is called the poynt blancke, for that vppon a leuell grounde wythoute anye hylles, as vppon the ſea, that
all thinges ſtandeth ſo leuell, ſhall bee ryghte wythe the
Horizon, that is to ſay, at the parting of the earthe and
the Skye, by the ſighte of youre eye : and then puttyng
downe the tayle of the peece A, the plummet line G wyll
                                                  hange

Figure 1.1  From W. Bourne, *The Art of Shooting in Great Ordnaunce,* 1578 (in
Tebeaux 1997, p. 217)

80          The Arte of

How to proyne or cut a Vyne in Winter.

This figure sheweth, how all Vynes should be proined and cutte, in a conuenient time after Christmas, that when ye cut them, ye shall leaue his braunches very thynne, as ye see by this fygure : ye shal neuer leaue aboue two, or three leaders at the heade of any principall braunch ye must also cut them of in the mpost betwéene the knots of the yung cions, for those be the leaders which will bring the grape, the rest & order ye shall vnderstand as foloweth.

Of the Vyne and Grape.

Somewhat I intende to speake of the ordering of the Vine & grape, to plant or set the Vyne: the plants or sets which be gathered from the vine (& so planted ) are best, they must not be olde gathered, nor lie long vnplanted after they be cutte, for then they wyll sone gather corruption, and when ye do gather your plantes, ye must take héede to
cut

Figure 1.2  From L. Mascall, *A Book of the Arte and Maner: How to Plante and Graffe all Sorts of Trees*, 1575 (in Tebeaux 1997, p. 21)

While Tebeaux claims that technical writing is as much a product of the Renaissance as is Dante's *Divine Comedy* and Giotto's "The Last Judgment", she admits that there are earlier writings in English that could be called technical writing:

"Chaucer's *Treatise on the Astrolabe*, written in 1391, exemplifies the best and perhaps the first English technical description ... [illustrating] several qualities that would surface repeatedly in technical descriptions of the sixteenth century". (Tebeaux 1997, p. 184)

Tebeaux goes on to state that:

"Chaucer establishes the tradition of describing a mechanism before presenting instructions for operating it, a method used in modern instruction and procedure manuals. Chaucer's method of integrating text and visuals and proceeding to describe the astrolabe according to the spatial arrangement of parts also fully anticipates modern practices." (Tebeaux 1997, p. 185)

But it is in the Renaissance that we begin to see the appearance of four features that are still valued today, namely:

- structured formatting and page design
- an awareness of the needs and abilities of the audience
- a plain, utilitarian prose style and
- augmenting text with graphics. (Tebeaux 1997, p. 133)

## What do technical writers produce nowadays?

Tebeaux's book is rich in examples of technical writing from the past, but what do technical writers write today? Content-wise, there seems no limit: manuals on the use of calculators, software (of all domains) and medical appliances; instructions for employees wanting to know how to apply for recreation leave; policies and procedures to govern employees' use of social media or access to sensitive documents; and so on and so on. And the sorts of documents that technical writers write are numerous:

- user guides (also known as user manuals, instructions for use and operating manuals)
- reference guides (for example, data dictionaries)
- getting started guides
- installation guides
- maintenance manuals (also known as service manuals)

- release notes
- online help
- policy and procedures documents
- work instructions
- technical data sheets
- product texts (field labels, error messages, etc.)
- technical marketing text
- tutorials and training materials: movies and documents
- bids and quotes.

One type of practical, how-to document that many wish professional technical writers would write, but sadly don't, is cooking recipes!

## What about writing?

If our professional adjective — *technical* — is misleading, what about our professional noun: *writing*? It is incontrovertible that what most of us do most of the time is write. Therefore *writing* does seem to be an appropriate noun.

But technical writers do, and have always done, more than just writing. When we create a document template or a cascading style sheet, we are doing the work of graphic designers. (Likewise when we create a flowchart or illustration to describe a process or procedure.) When we add an index to the back of a user guide or to an online help system, we are doing the work of an indexer. When we create an element definition document (EDD), write an XML transform or specialise DITA, we are doing the work of ... And now the waters muddy. Some of us spend a good deal of our time doing things other than writing. The question arises, then: do we need to reconsider the noun in our title — *writing* — as we probably do the adjective: *technical*?

Ignoring *technical* for the moment, there seems to be three possible approaches:

- retain *technical writing* for whatever practices folk who call themselves technical writers do, whether or not it involves much writing

- restrict *technical writing* to those practices where writing is pre-eminent and invent a new term—say, *documentation technician*—for those who assist technical writers but don't have writing as a primary responsibility or
- look for a new name that covers all the practices that all folk who now call themselves technical writers do.

The first and third approaches seem to have been the ones we have adopted, with varying enthusiasm from time to time. Many of us are happy to retain *technical writing* for whatever activities lead to the production of how-to materials. Others have pushed for names that downplay writing so as to better connote, if not denote, the wider range of activities we engage in. And thus instead of *technical writer* we have *technical communicator, content provider, end-user assistance professional, information designer, documentation developer, documenter*, etc.

Would consistency be a help to our profession? A single name would, surely, help promote us as a block to industry, commerce and government.

To keep our name or change it? Let's consider some arguments from both sides.

## Does doing more than writing necessitate a name change?

First, let's be clear that practitioners of many professions do more than what is connoted or implied by a literal reading of their profession's name. Teachers do more than teach. They also act as playground monitors, sports-day referees, mentors, excursion leaders and curriculum designers. But they still call themselves *teachers*. Likewise, surgeons give consultations, fire-fighters rescue cats from trees, accountants give financial advice. And yet one sees no moves by surgeons, fire-fighters and accountants to change the name of their profession.

Perhaps it is a question of how much time we spend doing things other than writing? If we spent 60% of our time on graphic design, then maybe we would be a graphic designer who also did some technical writing. Vice versa if we spent

60% of our time doing technical writing. We would be a writer who did some graphic design.

But what if we do a number of other things — product research, project management, graphic design, scripting macros, indexing and so on — that *together* occupy us for more than 50% of our time, with writing consuming the remaining time? Might not this justify a change of name?

But another sort of writer — a novelist, biographer or historian, for example — might likewise spend more time doing things other than writing. They might, for example, spend three years researching a book and only one year writing it. Should they be classified as, say, a *researcher* rather than as a *writer* on the basis of the relative times spent on each activity? I doubt it. It seems that what distinguishes them as writers, rather than as researchers, is the final goal of their various activities: to *write* a book.

The parallel is this: perhaps it doesn't matter how much time a technical writer spends on template design, illustrations, macro-coding, structured content rules, indexing and so on. If the goal of all these activities, taken together, is to produce a piece of *written* work, then they are *writers*. Just as Martin Amis is a writer regardless of how much research goes into his novels, a technical writer is a writer regardless of how much supporting but non-writing activity goes into the preparation of a user guide, online help system or the like.

## But some of us don't do any writing at all

We need to be careful not to beg the question here (in the sense of assuming what we want to prove). If some who call themselves technical writers don't do any writing, might one legitimate response be to ask what right they have to call themselves technical writers? Just as we might object to someone who paints houses calling themselves an artist, might we not object to, say, an illustrator or cartoonist calling themselves a technical writer?

But the work of some illustrators and cartoonists clearly resembles the work of a technical writer *when considered from the perspective of the goal of the activity.* If someone spends all their working time creating non-verbal illustrations that explain *how to do things—* such as how airline passengers should respond to an emergency—then this is clearly in the domain of technical writing as earlier defined: "the dissemination of practical knowledge (technical or otherwise), that is, knowledge about how to do things".

Here, then, might be a good argument for looking for an alternative to *writing* in *technical writing.* The *purpose* or *goal* of technical writing is a much more solid ground for defining what we do than the means by which we do it. There can be constancy in the purpose or goal without limiting ourselves to changeable means. In other words, the means can vary— writing, movie-making, illustration, and so on—while the goal remains the same: disseminating practical knowledge by means of instructions. Thus a better approach to naming our profession lies, perhaps, in finding a term that closely matches the profession's goal or purpose.

Before we explore this, note that some who work in our profession do not have, as their primary goal, the dissemination of practical knowledge. For example, some spend all their time creating element definition documents (EDDs), writing XML transforms, designing templates, coding VBA macros, specialising DITA and the like. These roles are akin to that of a laboratory technician, who provides the infrastructure for scientists to do their work but doesn't engage in any science. If we adopt a goal-based definition of our profession, then these folk may well need to be excluded. Perhaps a more suitable name for them is *documentation technician.*

## Does having a common goal necessitate a name change?

Focusing on our primary goal appears to be a good starting point for assessing what we should call our profession. But why must a shared goal—a goal shared, say, by writers and illustrators—require us to assume or concoct a name that in some way implies inclusivity for all who share that goal? A physician, osteopath and chiropractor all share a common goal: to make or keep people healthy. But physicians, osteopaths and chiropractors get by quite well without a common name, a name that somehow implies or connotes the activities of each and every such profession. Likewise with train drivers and bus drivers. Their common goal is to transport passengers to where they want to go, but no-one in their profession appears keen on changing these job titles to a name that in some way covers both activities.

In a similar vein, there seems to be no logical bar to having separate names for those whose shared goal is to disseminate practical knowledge *but who do it in different ways*, say:

- *instructional writer*: a person who spends most of their working time preparing written instructions, being what most of us in the profession do most of the time
- *instructional illustrator*: a person who spends most of their working time preparing cartoons, diagrams, illustrations and the like that give non-verbal instructions
- *instructional documentary-maker*: a person who spends most of their working time preparing animated instructions.

During the course of their careers, some technical writers will fall into two or three of these categories. But here the focus is on those who don't do any writing. It is the work of these folk who have prompted many a call to move our name away from *technical writing*.

## So, should the status quo remain?

This paper has considered some of the arguments put forward for why we should change the name of our profession. Those based on the fact that many of us do more than just write and that some of us don't write at all are not especially strong. Other professions have not felt a need to change their name because some practitioners do things other than what is implied by their name, and other professions have not felt a need to amalgamate under the one rubric just because they share a primary goal.

But that's no bar to us changing our name. The lack of a *need* to do something doesn't imply that we shouldn't do it. Indeed, what would prompt us to seriously reconsider our name is if we could come up with one that does neatly and inclusively capture what we primarily do: disseminate practical knowledge. Let's consider some of the terms that technical writers have called themselves of late and see if any meet this challenge.

# What have technical writers called themselves?

We have established that the common, longstanding name for our profession—*technical writing*—is inadequate. What we do need not be technical, and the purpose or goal of our profession can be met in ways other than by writing. This realisation has led some practitioners to adopt new names.

## Technical communicator

After *technical writer*, *technical communicator* is the most frequently used name for our profession. But it too has its flaws. For a start, the adjective is misleading. The domains we write about are often non-technical.

The noun too is not ideal. Consider a broadcaster of a radio show to do with the sciences. Such a person is called a *science communicator*: they communicate with the general public by informing them about issues of science. But a technical communicator is not someone who communicates with the general public by informing them about issues of technology. We deliver *practical* knowledge, *how-to* knowledge, not information of a general nature (as might a technology journalist). So, perhaps the term *communication* is too wide.

Also, *communication* has been appropriated by the spin industries. A job for a *communication consultant* these days is invariably a job for a public relations person or spin doctor. We would not want the profession seen by non-members as that of technology evangelists, spinning the benefits of technology just as a PR person spins the benefits of, say, a high-fat diet.

Perhaps it is for these reasons that many technical writers have sought to call themselves by names not obviously connected with *technology* or with *communication*. Let's consider some of these names.

## Content provider

This term is just far too broad. A journalist, graphic designer and musician can all be seen as content providers, so *content provider* hardly identifies and differentiates what we do in our profession from day to day.

## End-user assistance professional

This term likewise is far too wide. A call-centre representative is also an end-user assistance professional, as is a roadside breakdown mechanic, a librarian and a golf instructor. So the term doesn't identify and differentiate what we do.

## Information designer

Information is knowledge. To call ourselves *information designers* is to fail to distinguish between the sort of knowledge that is our prime concern from other sorts of knowledge. Technical writers are primarily concerned with imparting *practical* knowledge (that is, *how-to* or *procedural* knowledge), not with the sort of knowledge that, say, a physics teacher might be charged with imparting. A teacher who designs a physics curriculum is also an information designer; thus *information designer* is too wide a term to identify and differentiate what we do.

## Documenter

Not surprisingly, a *documenter* is someone who documents. If I document the species and number of birds that visit my backyard each year, I am documenting something: but I am hardly engaging in technical writing. Again, the term is far too broad to be a suitable substitute for *technical writer*.

## Documentation developer

This is the term preferred in the recent set of ISO/IEC standards. But the word *documents* applies to much more than the products that our profession typically develops. A diarist documents when they make entries in a diary; a public servant develops documents when they create a land title or a birth certificate. Documents are tendered in courts of law every day, but user guides are not.

# A survey of possible names

In 2009 I created a web-based survey designed to gather the thoughts of practising technical writers worldwide on what their profession should be called. The survey attracted 165 responses and 25 titles for our profession. Although the preamble to the survey included all the arguments I put forward above for suggesting that *technical writer* was a title in need of replacement, most respondents preferred it to anything else they could think of. In order of popularity, the titles suggested were:
- technical writer, 50 out of 165 (30%)
- technical communicator, 37 (22%)
- information designer, 24 (14.5%)
- technical author, 13 (8%)
- information developer, 7 (4%)
- documentation developer, 6 (3.5%)
- documentation specialist, 4 (2.5%)
- instructional writer, 4 (2.5%)
- content developer, 3 (2%)
- documenter, 2 (1%)
- communications specialist, 1 (<1%)
- content delivery architect, 1 (<1%)
- content provider, 1 (<1%)
- content specialist, 1 (<1%)

- content writer, 1 (<1%)
- developmental editor, 1 (<1%)
- document specialist, 1 (<1%)
- information engineer, 1 (<1%)
- information specialist, 1 (<1%)
- information technician, 1 (<1%)
- information architect, 1 (<1%)
- technical communication professional, 1 (<1%)
- technical journalist, 1 (<1%)
- user assistance and language expert, 1 (<1%)
- user support designer, 1 (<1%)

The survey results were also analysed by country, employment type (contractor or staff) and years of experience. These analyses are given in appendix A starting on page 239.

## Some humorous suggestions

The survey also called for humorous titles that befit our profession. Here are some of the responses:

| | |
|---|---|
| blarney interpreter | information tamer |
| bull dispersion operative | jargon interpreter |
| clarity consultant | language hardhat |
| communication acrobat | manual labourers |
| communication carer | meaning sculptors |
| communication communicator | message maestro |
| de-complicator | professional explainer |
| demystifier | semantic engineer |
| disambiguation engineer | sensemaster |
| doculologist | sentence constructionist |
| documentationalist | specification deboggler |
| geek-speak interpreter | word engineer |
| gobble-de-gook translator | word monkey |
| ignorance obliterator | wordsnotworth |
| information distiller | wordwrangler |

# What do most of us do?

The survey also asked what activities take up most of your time. This was an optional question, but with a purpose. It was designed to validate, or otherwise, a gathering belief that writing and editing are becoming minority activities in technical writing. The survey results suggest that this is not the case.

Of the 165 respondents overall, 58 chose not to answer this question. Of the remaining 107, 85 (or 79%) indicated that *writing* and/or *editing* are the activities that take up most of their working time. This leaves 22 (or 21%) who spend most of their working time doing things other than writing or editing. (Some in this latter group stated that they were team leaders and led teams of folk engaged in writing and editing. Others said that they spend most of their time researching, which presumably means, in many cases, researching topics that they need to *write* about. Hence the proportion of those who are primarily involved in writing and editing is likely to be greater than what is suggested by the raw survey statistics.)

Our conclusion must be that writing and editing still constitute the activities that take up most of the time of technical writers.

## What else do we do?

The following is a list of what activities survey respondents do in their professional life other than writing:

- researching
- fact-checking
- planning
- managing (or team-leading)
- project and people management
- instructional design
- content design
- teaching
- illustrating (and graphic design)
- interviewing
- information design

- content management
- translation support
- coaching or mentoring junior writers
- negotiating
- template design
- photography
- GUI design (and review of GUI designs).

## A sprinkling of respondents' comments

"I think a name change is warranted because of the changing nature of the software industry. At one time, Technical Writer was the most appropriate name. However, as our skills are increasingly being applied to UI design and usability, and as printed manuals start to die out, I think it's time to let go of the 'writer' word. I believe we are experts in providing appropriate assistance. We do this using our knowledge of language and writing. It is important that we are considered by management whenever either of these (assistance or language) are in question."

"Please, please, please do not suggest made-up words like 'documenter' or, even worse, a term I heard floated at one company where I worked—'documentationist'! This survey is rather silly if you ask me."

"[I prefer] *instructional writer* because it's slightly closer to reality for those of us who don't always write software manuals. Also, it might require slightly less explaining to acquaintances and neighbours. 'I'm a technical writer. I write and edit instruction manuals and web pages' just takes too long."

"I have found that 'technical writer' is the most easily understood label for our profession. Thanks for running this survey, it has stimulated a lot of debate on techwr-l and elsewhere."

"Hopeless task ... you will never get consensus."

"In spite of calling myself a communication consultant on my business card, I still think of myself as a tech writer and that's what I believe the profession should be called forever. Let's not try to confuse the world."

"Keep it simple with the name that has been around and is more or less understood. Changing it—even if more descriptive/accurate—will only create confusion and more explaining. When I am at a party I say 'I don't write software, I write *about* software' and people say 'Oh, you're a technical writer, and you do user manuals?' Yep. And marketing copy and web site and other stuff. They get it."

"'Designing' info is a bit pretentious, isn't it, but often the design part of the job takes up as much, if not more, time than actually writing or editing it."

"Do *not* change the name. We deal with words and those words have to be written. We are writers. 'Technical' doesn't have to mean engineering: think of legal cases where someone gets off on a 'technicality'. When we talk about technical writing, we mean precision and accuracy. Poets can afford to be vague and leave the interpretation to the reader; technical writers cannot. The phrase just means that the words you use must be unambiguous, clear and concise. Therefore, whether you are explaining how to appeal against a parking ticket, describing how to change the brushes on an electric motor, or providing practical advice on achieving creative results with watercolours, what you are doing is technical writing. Learn to live with this."

"It surprises me that this debate crops up so often in a profession that supposedly rejects jargon. 'Technical author/writer' is what we're called, whatever the job entails. Companies understand this term. If you use anything else you're diluting the 'brand' and using jargon. Perhaps we should call teachers 'learning facilitators' or doctors 'wellness promoters'. Perhaps we should grasp more firmly a term that is already commonly used so that we can get more recognition as a profession."

"The problem with finding a term for our profession is that technical communication is such a broad and varied stripe. Not all tech comms is instructional, not all is help development or end-user documentation or even written documentation. That's why we can't agree on a term that both fits the profession and improves its image."

"I don't think that technical writing/comms etc is a discrete profession any more, and it's all a bit People's Front of Judea versus Judean People's Front (to misquote *Life of Brian*). The issue for us is to get people to see that we are communicators, technical or otherwise."

"I've been called a Technical Author in England, plus I've been called an Information Developer, and now a Technical Writer. I've also been a Documentation Manager and a Technical Publications Manager. It doesn't really matter what we're called as long as we're employed and make a difference to the organisations that employ us through the value of our publications. When we stop adding value through our publications we won't get employed and we'll be called *unemployed*. Technical Writers/Technical Authors/Information Designers/Information Developers have resisted certification/registration that most professionals have. I've recently studied and qualified as a Network Administrator (CCNA) to help me write networking documentation in my current organisation, and have found the IT profession is now very heavily certified. The IT networking staff I work with are universally certified whether it's Cisco CCNA or Microsoft MSCE."

"Many of the technical writers I've worked with over the years are neither *technical* nor *writers*, but prefer to work as desktop publishers. I think companies frequently see the two as the same, and they are frequently shocked when a true technical writer gets involved."

"I am quite happy being a technical author, mainly because it amuses me to see the blank looks when asked what I do for a living!"

"It depends on what mood I am in and how many times I have been stonewalled by a developer or programmer, but the term 'Glorified Secretary' or 'Glorified Typist' sometimes makes me laugh and sometimes makes me want to shout at people that technical writers are not just typists, that research and asking questions are more part of the job than the actual writing."

"I'm not in favor of the current push to call those of us in the profession *technical communicator* on the theory that technical writer doesn't describe everything we do. Well, my perspective is that technical writer is so entrenched in the marketplace that it's not going to go away and we're not going to be able to change it."

"The term *technical writer* is established in the industry. What you need is a robust glossary definition of the term. Why not ask for that?"

"Changing names at every subtle change of the main theme of our work is just plain silly. I am still having to explain *IT* and *ICT* to people. I tell them it is a silly name for *computing* and they are happy."

"I think a variety of names reflecting similar skills is probably the best. There are a variety of different jobs performed by people in our profession. My resume would probably not qualify me for the types of jobs performed by others who belong to the same professional organisations as I do. I also think this largely reflects the state of play in other professions: very few professions have names which match with universal functions."

"I like the existing, long-standing name and believe that it probably just needs to be marketed better. I like the idea that technical writing is a profession that belongs in many industries and that, like many professions, technical writers will probably do less writing and more management as their careers unfold. No to a name change!"

# 2: Ethics and technical writing

Ethics—or *moral philosophy* as it is also called—has been the subject of philosophical investigation for thousands of years. What should we do and not do? What sort of person should I be? How should I live? These are questions that have exercised some of the greatest minds: from Aristotle (and no doubt earlier) to Peter Singer (and no doubt later). Their search has been for a general principle that reflects our strongest moral intuitions, one that might guide us in our quest to lead a moral life, a good life. And this is no trivial matter. Whether we acknowledge it or not, morality influences many of our personal decisions. It also drives the formation of many of a society's laws.

Sometimes the moral dimension of our actions is not clear, drowned out by custom, greed, laziness or simple ignorance. It can take much effort, argument and even radicalism to bring a society to see the immorality of hitherto accepted actions or customs. For example, slavery, animal cruelty and disenfranchisement based on gender were once deemed perfectly acceptable—until an evolution in moral discernment drew out their inherent repulsiveness. That evolution is still underway, with more and more of our actions and customs coming under the moral microscope for the first time. One question such an evolution invites is whether the force field of morality extends also to technical writing?

Fist published in *Southern Communicator*, edition 31, February 2014. Addenda added in 2015.

A core notion in ethics is the undesirability of doing *harm* to others (where *harm* is understood as physical harm, mental harm or being made worse off). Any code of conduct that had nothing to say about such actions could not, by definition, be classified as a system of ethics. (A personal-fitness code of conduct would fit into that category.) Understood as a system of harm-mitigation, morality does, as I'll argue below, have relevance to technical writing. Moreover, it imposes obligations not just on technical writers, but also on those manufacturers who produce products that technical writers typically write about.

## Ethics and the technical writer

First, the technical writer. There are at least four measures of the moral suitability of the end-user documentation that technical writers typically produce:

- Does it pose any risk of physical harm?
- Does it pose any risk of mental harm?
- Does it offer what is reasonably expected of it?
- Does it impose unreasonable costs on readers?

Here are some examples of potential physical harm that clearly show that end-user documentation can come within the force field of morality:

- Poorly written instructions could be what turns a recoverable mid-air emergency into a disastrous plane crash.
- Poorly written instructions could injure those who operate or maintain dangerous equipment (such as lathes, grizzly feeders, mains transformers, and so on).
- A service technician, working from poorly written instructions, might mis-calibrate a diagnostic device, subsequently leading to mistaken diagnoses, ineffective treatment, pain or even death.

Second, the risk of mental harm. A person's self-esteem, and general mental wellbeing, can be affected by the degree of confidence they have in their ability to understand the written material they need to understand. For example, a young apprentice having difficulty understanding operating instructions may give up their chosen trade because of a mistaken belief that they do not have the intelligence to succeed in it. Mistaken? If their difficulty arises not from poor language skills on their part but from carelessness on the part of the writer of those instructions, then the apprentice's belief that they are not up to it might very well be mistaken. Their chosen career might be denied them by the thoughtlessness or laziness of writers who have provided them with vague or impenetrable instructions.

Third, the notion of "my station and its duties", introduced by the British philosopher F. H. Bradley, is central to morality (Bradley 1876, essay 6). The complexity of modern life means that we cannot do all that we might want to do. Thus we employ others to do it for us and, more importantly, *trust* that they will do it. We entrust teachers to teach and nurses to nurse, and there is a strong societal expectation that they will do just that (for without trust a society quickly collapses). If you accept the station of nursing, you accept the burden of others' legitimate expectation that you *will* look after and comfort the sick. Failure to do so deserves moral reproach. If you accept the station of teacher, you deserve moral reproach if you fail to teach. Likewise, if you accept the station of documenter, you accept the trust that others have placed in you by documenting what is expected. You are under no less a moral obligation to instruct than a teacher is to teach. To deny that is to deny the moral relevance of trust.

Fourth, time is everyone's most valuable asset. It might be an intangible asset, but it is clearly more valuable than any tangible asset. (What good is having a Ferrari if you haven't the time to drive it?) Time is a birth right. And like many other intangible assets—goodwill, patents, copyright, licenses, and so on—a price can be put on time. For example, your salary is the price you accept for selling your time to others. Now if the theft of a tangible asset—such as a car—is a moral issue, then so must be the theft of an even more valuable asset: time.

Hence readers are right to feel aggrieved at having to swim through the time-stealing treacle of verbosity, redundancy, tautology, padding, nominalisation, procedural bloat, and so on to get the information they need. They are having time stolen from them that could be put to more profitable use. So technical writers have a moral obligation to write in ways that ensure *efficient* communication, communication that does not steal their readers' time.

## Ethics and the manufacturer of goods

What now of manufacturers? There is little doubt that the relative quantity and quality of end-user documentation has fallen over the last few decades. When once a comprehensive user guide accompanied every product, now it is just as likely that all that will be provided is a quick start guide, no documentation—as with many smartphone and tablet apps— documentation that is difficult or costly to access, or documentation that is barely useful. Is this reduction in instructional offerings a legitimate commercial strategy? Or are there countervailing moral considerations?

Consider the following cases:
1. You buy an electric kettle but there is no power cord in the packaging.
2. You buy a portable radio and there are no batteries in the packaging.
3. You buy software and the instructions are only available while you are connected to the internet.
4. You buy software and there are no instructions on how to use it.

In each case, many if not most purchasers could not use the product as it is. What the product is advertised as being able to do cannot readily be done. Something more is needed. In other words, the product is not *fit for purpose* as it is. In case 1, the omission was no doubt an oversight. A retailer keen to remain in business would ungrudgingly give the customer a cord. In case 2, the omission was probably intentional. If not used for some time, batteries lose power and are prone to leak. So they are often excluded from shelf-bound products that need them. But in this case, ethics—usually supported by consumer law— insists that purchasers are made aware beforehand that further expense will be required. This usually takes the form of an indicative marking—such as "Batteries not included"— displayed prominently on the packaging. For consumers have a right to know the *full* costs of what they are buying: the radio *plus the batteries*. It is a right grounded in the value *Homo sapiens* place on honesty. It is simply dishonest to sell a product for $x knowing that the true cost to the purchaser will be $(x + y)$. And honesty is a fundamental moral expectation, to be dispensed with only in extreme circumstances. If you are unconvinced, consider this parallel: you buy a car from someone for $4000 not knowing what the seller knows, namely, that another $4000 will need to be spent on the car to make it roadworthy. Has the seller acted honestly in not telling you? Has the seller acted morally?

Now just as the purchaser of a battery-powered radio has to spend more to get the radio to work, the purchaser of, say, software sold without a user guide—or with instructions that are only available while the user is connected to the internet— will have to spend more to get the product to work as advertised. Tangible assets (money for books or internet access) and intangible assets (time) add to the cost of the software just as batteries add to the cost of the radio. The latter calls for a warning on the packaging; why not the former? If the seller of a car behaves immorally in not informing you that the car needs more money spent on it to make it properly usable, isn't a company that sells you *any* product knowing that you will need to spend more to be able to properly use it also behaving immorally?

It might be argued that:

- it's acceptable that goods are fit for purpose *for some customers*
- the omission of user documentation has lowered the price of the product (which, presumably, is of greater utility than having documentation readily at hand).

It is true that producing products that are fit for purpose is not an *unqualified* obligation on manufacturers. The manufacturer, say, of a game suitable only for those over 10 years of age has not produced a product unfit for purpose just because a toddler cannot understand it. So perhaps software that *some* people can use unaided is not necessarily unfit for purpose. But the manufacturer of the game is obliged to prominently place an age marking on the game's packaging (such as "suitable for children over the age of 10"). For it would be *dishonest* to take money from parents for games their children would not understand. Perhaps, then, the manufacturer of software should be obliged to affix a marking indicating that it is not suitable to all potential purchasers. A toddler has to wait to become more intelligent before being able to appreciate the game; the non-savvy software purchaser has to wait to become more informed before being able to use the software. In the former case, manufacturers are obliged to alert purchasers; why not in the latter case?

Age or skill-level markings are not needed on all products. Caterpillar, for instance, doesn't have to display markings on their tractors to indicate that they are suitable only for farmers. But that's because Caterpillar only markets tractors to farmers. Why market tractors to school students? But it is different when a product becomes cheap enough for most of us to buy and it is marketed without demographic discrimination. Take computers for example. They are now so cheap that just about anyone in a first-world country can afford to buy one. Moreover, they are marketed to everyone. And that includes many who have not had much prior exposure to computers: those whose previous occupations never required them to master a computer; parents returning to the workforce after a long absence; senior citizens. These people deserve to know that they will not, *without further expense,* be able to make the product do what the magnet of marketing convinced them it could do.

It might be retorted that consumers should expect to have to spend some time learning how to use a new product. Yes, but the impost on their time should be *reasonable*. And what is reasonable is easy to determine. You take the entire demographic to which the product is being marketed and consider the person likely to have the *least* background knowledge about it. The time it would take this imaginary person to work through and apply a well-written procedure is reasonable. If it would take longer — because the procedure does not exist, is poorly written or is buried in the archives of some obscure forum — then the call on some customers' time will be unreasonable. It doesn't matter that some customers will learn the task more quickly. The point is that both sub-demographics — the less knowledgeable and the more knowledgeable — are being marketed to. If a manufacturer is going to take assets from *any* customer, it is only fair that the customer knows what those assets are likely to be.

So honesty and fairness impose on manufacturers an obligation to provide consumers with usable instructions or clearly label their product packaging in a way that makes the true cost of purchase estimable by those to whom the product is marketed. Package-markings like the following would help manufacturers meet this obligation:

- ☐ Instructions included
- ☐ Instructions available by one-off internet download
- ☐ Internet connection needed to access instructions

where "instructions" means instructions on how to use each advertised feature and each feature a consumer can reasonably expect the product to have. Only then are consumers given an opportunity to make an *informed* decision about the true cost of their intending purchase.

So far we have only considered whether consumers have a moral right to be informed of the presence or absence of instructions and the potential cost in retrieving them. Might consumers have an even stronger moral right, namely, to expect that there *are* instructions on how to use a product?

First, consider whether consumers might be willing to give up that right if the omission of instructions brought about a significant reduction in the retail price of the product. To claim that they necessarily would assumes that the retail price is the only measure of cost to the purchaser, and hence the cheaper the better. But this is blatantly wrong. The true cost of any product includes various externalities. For example, the true cost of electricity includes the cost of environmental degradation (and far exceeds the cumulative charges that appear on consumers' electricity bills). Likewise, the true cost of a product that one wants to use must include the opportunity cost of the time taken to learn how to use it. If it takes me 2 minutes to find out how to do something by reading about it in a user guide, but 12 minutes by trial and error or forum-trawling, then 10 minutes of my time has out-competed 10 minutes of time that could have been put to purposes that yielded me greater utility. Suppose there are 15 things I need help with, each of which involves trial and error or forum-trawling. All up, 150 minutes of my time will have been spent on tasks that could have been substituted with tasks more pleasurable or profitable. You might argue that by undertaking that learning I have judged it to be the most valuable activity to me at the time, otherwise I would do something else. But is the impost on my time fair? Based on the average salary in Australia, that amount of trial and error and forum-trawling equates to approximately $100 of potential income. Does a user manual add $100 to the cost of a product? Hardly. Amortised over, say, 50,000 sellable units, a 200-page printed user manual with a usable index would add no more than $10 to the cost of the product (or $2 if provided only in electronic form). So your net loss is at least $90 (and more if you earn an above-average salary). Thus cheaper in the store does not necessarily equate to less costly overall.

But even if it did lead to net savings to consumers, omitting instructions from some products should set the moral Geiger counter ticking. It is true that the use of some products is too intuitive to warrant instructions: pens, and books for example. But what of non-intuitive products marketed to everyone indiscriminately? Computers and mobile phones fall into this category. Given the bell curve of general intelligence, it is clear that some purchasers of such products will not be able to get them to work as advertised without instructions. (Recall that their use is non-intuitive, and note too that 50% of people are, by definition, below average intelligence.) To sell a product knowing that some purchasers will not be able to use it is simply deception. It is tantamount to selling a product knowing that perhaps as many as 50% of units don't work as advertised—and with a no-returns policy. It is taking money under false pretences, the immorality of which should be obvious. And it would still be so regardless of the percentage of purchasers who couldn't use the product. Hence providing instructions—or offering full refunds—is a moral imperative in such circumstances.

Poor or absent documentation can, then, fall within the force field of morality by it being the result of dishonesty and deception: the manufacturer fails to reveal the true cost of the accompanying product or sells a product that is unusable. It can also cause harm: asset theft will make some, perhaps all, consumers worse off than they reasonably expected to be. Thus the current diminution of instructional offerings that society seems to be acquiescing in is not a legitimate commercial strategy. Morality is on the side of consumers, not manufacturers. It is time for the law to catch up.

## Addendum 1: But no-one reads user guides, right?

This is a claim often made by managers keen on cutting costs, but it is patently wrong. It's true that most users of a product— whether it be a steam iron or a mobile phone—prefer to figure out for themselves how it works. They want to start using it as quickly as possible. They don't sit down and work through the user manual first. Most people know that *learning by doing* is more effective than *learning by reading*.

But that doesn't mean that they don't read user documentation. Rather, they dip into it, when needed.

When they do dip into the user documentation, it is usually when they are in the middle of doing something else. For example, they're in the middle of running the monthly payroll. The new version of the payroll system they have just installed is not as intuitive as they had hoped. They can't get the pay-slips to print, but they need to upload all the payroll data to the bank in the next 15 minutes. If they ring customer support, they'll probably be left hanging on the end of the phone for longer than they've got. So they will go to the user documentation—the user manual or the online help—and hopefully find the answer they need.

And here we have another reason why *good* technical writing is important. In most cases, our customer—the reader of our documentation—has come to it in annoyance and frustration. They've got better things to do than read about how to do something that they feel should have been intuitive. How much more annoyed and frustrated are they going to be if the user documentation they are consulting as a last resort is so poorly written that they cannot understand it.

## Addendum 2: Moral scepticism

We have argued that harm-minimisation lies at the heart of morality and that, by potentially harming users of products, technical documentation falls within the force field of morality. The moral sceptic might retort that all we've done is elevate mere feelings to a status they don't deserve, neglecting reason and logic in the process. Feelings are wishy-washy things, subjective and ever-changing. To base a system on mere feelings is to open the door to relativism. And the sceptic will be keen to point out that what is passionately considered to be immoral in one society can be considered amoral or even moral in another. Further, the sceptic might also be keen to remind us that philosophers have been attempting to ground ethics on reason and logic for at least 2500 years, and without much success. Perhaps, then, our sense that there are moral imperatives—legitimate    pressures    on    us    to    behave    in particular ways—is just a delusion, an atavistic feeling with no logical basis.

Let's consider these three retorts in turn. First, why must reason and logic trump feelings in all fields? We would, rightly, ignore feelings in determining the answer to a mathematical problem. And although it might prompt a scientific hypothesis, gut feeling also has no place in scientific experimentation. But when it comes to the desire to avoid fear, anxiety and harm—a desire so strong as to be instinctual—we cannot escape from the realm of feelings. For fear, anxiety and harm are expressed as or through feelings. Now since ethics is primarily concerned with the avoidance of fear, anxiety and harm, it cannot avoid a consideration of feelings. Ethics might also be concerned with character, but that doesn't detract from the fact that it is inextricably concerned with feeling. Further, reason and logic do have a place in moral reasoning even if the foundation of morality does lie in feelings. In attempting to determine whether, say, abortion, price-gouging or trading in pollution permits are immoral or amoral acts, ethicists employ the same forms of logical analysis one sees in many non-moral disciplines, forms such as analogy and implication. Such analyses might reveal, for example, that there is unrealised harm in some practice or other, or that a particular moral view-point is inconsistent, or it leads logically to conclusions that holders of that view might not accept. Like other forms of reasoning, moral reasoning can, in other words, be logical or illogical, rational or irrational.

Second, does moral relativism prove that morality is an illusion? In other words, does a difference in moral views between societies necessarily imply that there cannot be an objective moral view, one that might eventually apply to all societies? There are many people who believe that the earth was created in seven days around about six thousand years ago. There are many who disagree with this view, positing the origin of the earth (and everything else in the knowable universe) in the Big Bang. Does the disagreement between the two groups imply that neither view is correct? Likewise, many believe in anthropogenic climate change and many do not. Again, does the disagreement between the two groups imply that neither view is correct? The answer, of course, is no. Similarly, the fact that one society considers it morally obligatory to stone an adulterer while another finds it morally repugnant doesn't imply that neither view is groundless. It might well be the case that both views are groundless, but you do not prove that merely by showing that the views are in opposition. Indeed, it might well be that:

> "differences of opinion on moral questions merely reveal the incompleteness of our knowledge; they do not oblige us to respect a diversity of views indefinitely." (Harris 2010, p. 10)

Third, the failure to have discovered an overarching moral principle—a principle that gives ground to all the various forms of moral reasoning—is no reason to conclude that ethics has no basis, that it is a myth or a relic from the pre-scientific era. The fact that no overarching theory has been discovered that unifies the four fundamental forces of nature—gravity, electromagnetism, the weak nuclear and the strong nuclear forces—does not mean that these forces are not real. Just as gravity asserts its influence despite the lack of a unifying theory of forces, harm-minimisation could well drive some forms of moral reasoning despite the lack of a universal, overarching principle of morality. And just as scientists continue to strive towards a unifying theory of forces, there's no reason why ethicists should not continue to strive towards a universal, overarching principle of morality.

We should realise that the study of ethics has only recently shrugged off the specious influence of religion and may yet encounter its own Enlightenment (as the sciences did in the eighteenth century):

> "The Earth will remain inhabitable for at least another billion years. Civilization began only a few thousand years ago. If we do not destroy mankind, these few thousand years may be only a tiny fraction of the whole of

civilized human history ... Belief in God, or in many gods, prevented the free development of moral reasoning. Disbelief in God, openly admitted by a majority, is a very recent event, not yet completed. Because this event is so recent, Non-religious Ethics is at a very early stage. We cannot yet predict whether, as in Mathematics, we will all reach agreement. Since we cannot know how Ethics will develop, it is not irrational to have high hopes." (Parfitt 1984, p. 454)

But there is no need for mere "high hopes" when we can in fact be optimistic. For it is clear that moral development and evolution has indeed been occurring without the influence of religion, and in ways generally, if not universally, accepted. For example, as a result of the women's liberation movement that gained momentum in the 1960s, many societies now treat women with a respect once not accorded to them. This is an unmistakable advance in our moral attitudes that few would honestly deny. But it did not come about because of religion. It came about because of the logic and power of the moral reasoning behind it. The moral values of fairness and of harm-minimisation were seen, on examination, to far outweigh the paltry and corrupt rationalisations of male-dominated societies.

There have been other advances in moral thinking that have led to worthwhile practical outcomes. We now appreciate that animal suffering falls within the purview of morality, and governments—either by their own acceptance of the fact, or at the insistence of their electorates—have enacted laws to prevent (or at least minimise) animal cruelty. The acceptance of indigenous rights, of homosexual lifestyles, of the amoral nature and indeed naturalism of that morally loaded notion of "self-abuse"—the acceptance of these and many more represent advances in morality achieved without recourse to religion. Religious groups may have supported some of these developments, but the reasoning behind them was not religious. Rather, it was based on notions of fairness and harm, notions that lie at the very heart of morality.

The true moral sceptic—one who thinks morality is a
subjective delusion—is likely to be perplexed by moral
development and improvement. Why is life better now,
morally better, than it was, say, a thousand years ago? It is
difficult to doubt that life is better and that the development of
morality—whether you think it was spawned by religious
belief or otherwise—has made a significant contribution. But
why, if morality is entirely subjective and groundless, should
we have bothered to struggle, debate, demonstrate and even
immolate in the name of morality or in the hope of further
developing morality? Few, if any, now believe that human
slavery or the stoning of adulterers is morally justified. The
repugnance we—that is, the vast majority of people—feel
towards such practices is a moral repugnance. It is not a
delusion. It is, rather, a feeling that has gathered strength over
many centuries (sometimes as the result of much argument).
Our erstwhile barbaric acceptance of such practices has
transmogrified into a deep-seated feeling that has become
yoked to our reason and to our action. We now actively rebel
against such practices, acknowledging that the world is a far
better place for it being rid of the harm they inflicted. We have,
in other words, become better moral beings over the centuries.
While we cannot assert this generally, we can assert that, on
balance, life in the twenty-first century is less anxious, less
brutal, less unforgiving than it was a thousand years ago. One
is less likely these days to flee from marauding armies, or to be
left to die on the side of a road, or to be sent in chains to a
foreign country to break rock or pick tobacco day in and day
out. Why the change, and a change for the better? It is not
advances in religions—whatever that might mean—that has
effected such changes. Nor is it advances in science, for science
is morally neutral, capable of being put to good use or bad. Nor
is at, as some economists believe, the rise of free markets, for
the reach of such markets has always been restrained, often
imperfectly, by moral considerations. (Did free markets bail out
the morally and financially bankrupt banks during the Global
Financial Crisis of 2008?) Indeed, the notion of a wholly

unfettered market easily gives rise to moral qualms. A society where health care was totally in private hands, where the market sets the charges for medical care, could lead to immense suffering, especially among the poor. Further, would free markets ensure that every child, regardless of their parents' income, gets a decent education? And would free markets prevent trading in children if there was a profit to be made?)

Rather, it is advances in moral thinking, combined with the continual fine-tuning of our moral sensibilities, that has led to many of the improvements in living we now enjoy. Science and markets—and taxation—have indeed provided the material wealth many of us enjoy today, but the way we decide to share and distribute that wealth is not based on scientific or economic thinking. Nor is it based on religious reasoning. Rather, it is based on moral reasoning. A society that gives its citizens access to health care regardless of their assets or income is a society that has judged health care to be a moral right. The extent to which they can achieve the goal will be determined by the wealth available. But choosing that goal is not determined by how the wealth was achieved—whether by the application of scientific discovery or by market trading. It was determined by the thinking, arguing and persuasion of social activists, academics, politicians and others who had equality and harm-minimisation as guiding moral values.

Given that life is, in general, less nasty and brutish than it was a thousand years ago does not mean that we don't slide backwards on occasion. But it seems that for each step backwards there are two steps forwards. There is a trend line that should give us all cause for optimism, whether or not the current state of our moral optimism sees us below or above the line. Overall, our moral sensibilities are sharpening, and who knows what lies around the corner. Two hundred years ago, no-one had even conceived of quarks and bosons, but now they are fundamental to our understanding of the nature of matter. Two hundred years ago few had pondered the moral status of women's equality or of assisted suicide. Like scientific knowledge, moral knowledge too can evolve.

But we shouldn't expect too much. We might never reason our way to a single overarching moral principle that can account for all our moral intuitions and convictions, those to do with justice, equality, compassion, harm-minimisation and the like. But that doesn't necessarily mean there are no rational principles covering these facets of morality when considered in isolation. We could well accept that we are morally obliged to minimise harm even though the value of harm-minimisation cannot be derived from any higher principle. And it's no weakness in that view to accept that there may be exceptions and qualifications: you are morally obliged not to harm others unless ... Science too is replete with exceptions and qualifications: water boils at 100°C except if the water is not at sea level; light travels in a straight line except when subject to the gravitational pull of a large body; and so on. We do not abandon science because so many laws of nature are subject to exceptions and qualifications. Likewise, there is no reason to abandon the morality of minimising harm solely on the grounds that there may be exceptions and qualifications.

But let us, for the sake of argument, go along with the moral sceptic and accept that ethics is a subjective and ultimately groundless concern. Might not ethics still have a practical and worthwhile purpose? The same arguments that ethical sceptics propose could apply just as equally to law. Ultimately, there might be no logical foundation to any law, no unshakeable axiom or set of axioms from which all laws can be derived. But would that be a reason for societies to abandon law? To borrow an idea, and a well-worn phrase, from Thomas Hobbes, a life unregulated by laws would soon become "solitary, poor, nasty, brutish, and short". To the moral sceptics reading this: would you be happy living in a society where there were no laws?

Although they are clearly different, there is much overlap between morality and law. Some issues of law are outside the scope of morality: driving an unregistered car and jay-walking, for example. Similarly, some issues of morality are, in most countries, outside the scope of law, such as infidelity, miserliness and plagiarism. But laws proscribing murder, rape, theft, deception, false imprisonment and many other actions are firmly rooted in moral conviction. In other words, law owes much to morality. To see this, imagine a society that abandons the law forbidding murder. There is little doubt that that society would still retain a strong moral proscription against murder. (The aversion to death is strongly rooted in *Homo sapiens*, and no doubt most if not all animals.)There is no doubt too that, in order to deter others from action that we are so strongly averse to, murderers would continue to be punished. The punishment will take different forms—banishment, ostracism, private retribution and so on—rather than those currently sanctioned by law. But whatever the form, its existence would clearly indicate a societal need. So even if proven ultimately groundless, morality (and its associated mores) would still be invaluable in helping societies avoid the Hobbesian jungle. Law, in large part, is based on morality. So if morality is groundless, then so too is much (perhaps all) of law. And if law is what we need to avoid the Hobbesian jungle, then we likewise need morality.

This doesn't mean that law is always a perfect reflection of some aspect of morality. To borrow from Charles Dickens, the law can be an ass. It can be poorly thought out, and it can be usurped by powerful vested interests. As an example of a poorly thought out law, consider the law enacted in some states in Australia that declared that anyone who sends child pornography via electronic means is guilty of an offence and their name must be added to a publicly accessible sex offenders register. The law intended to limit the activities of paedophiles, a laudable goal, you might think. But it inadvertently caught in its net under-age teenagers pranking around sending nude pictures of themselves to their friends. Some youngsters found them selves listed on the sex offenders register, forever tarred as a sex offender and thereby denied many rights enjoyed by others (such as the right to be a school teacher). Clearly, such an ill-thought-out law is an ass. A law that gives preferential funding to private schools over public schools could legitimately be seen as one mid-wifed by politicians beholden to vested interests. Morality and law can also be in conflict in other ways. A surgeon who exceeds the speed limit in rushing to a hospital might have broken the law. But if, in so doing, the surgeon manages to save the life of a seriously ill patient, such speeding would be an act deserving of moral commendation. Even if the law made provision for judicial discretion, there is no guarantee that it would be applied morally in any particular case. That is the very essence of discretion.

So even if it is largely based on morality, law is not necessarily moral.

The domain of the law is continually expanding to cover areas once the sole domain of morality. The strong moral convictions of William Wilberforce, and the arguments he put forward, led to the legal abolition of the slave trade in Britain. Strong moral conviction, and strong arguments, have led to laws being introduced in many countries to protect the welfare of animals. Strong moral conviction, and strong arguments, have led to laws being introduced in many countries aimed at avoiding catastrophic climate change. In other words, much that is now in the purview of law was once solely a matter of morality (again proving that law often grows out of, and is based on, morality).

Thus, to abandon morality is, in large part, to abandon law. So, to the moral sceptic, we ask once again: would you prefer to live in a society where there are no laws?

These considerations do, I trust, prove that morality is worthy of respect, that it is intrinsically linked to harm minimisation and that technical documentation, by its ability to harm readers, comes within the force field of morality. And just as considerations of morality have been increasingly replicated in law, so too has the morality associated with documentation (technical or otherwise). Organisations have been successfully sued by plaintiffs claiming that they have been disadvantaged by an inability to understand certain public documents.[1] To protect themselves against such litigation, many bodies—

commercial and government—now have specific guidelines on the minimum level of readability required of their public documents. For example, numerous US insurance companies demand this of their customer insurance policies, and the US Federal Drug Administration demands a readability level no greater than that to be expected of an eighth-grader on medicinal labels and on medical consent forms. The National Cancer Institute and the Office of Human Research Protection (both in the US) make similar demands on literature destined for public consumption.

The insistence that writers in the American civil service adopt simple, utilitarian, plain English had been gathering strength since the presidency of Bill Clinton. In 1999, Clinton issued a Presidential Memorandum declaring that all government regulations should now be written in plain English. Clinton's vice president Al Gore famously declared that "plain language is a civil right". During the George W. Bush presidency, The *Plain Language in Government Communications Act 2008* passed through the US House of Representatives, the purpose of which was:

> "To enhance citizen access to Government information and services by establishing plain language as the standard style for Government documents issued to the public."[2]

On October 13 2010, President Barak Obama strengthened plain-English requirements with the *Plain Writing Act of 2010*, the purpose of which is:

> "to improve the effectiveness and accountability of Federal agencies to the public by promoting clear Government communication that the public can understand and use."[3]

---

1. For example, Tampa General Hospital and the University of South Florida paid a US$3.8 million settlement to a group of people who claimed that a consent form they signed exceeded their reading ability. Cited in DuBay 2007, p. 2.
2. See http://www.ssa.gov/legislation/legis_bulletin_041408.html. Viewed 22 September 2015.
3. See https://www.fdic.gov/plainlanguage/plainwritingact.pdf, section 2. Viewed 12 September 2015.

Thus the law is catching up, if only sporadically, with the moral imperative to write in an audience-centric way if the goal is to impart information. And such a style of writing is the primary hallmark of good technical writing.

# 3: The cost of poor writing

Some folks—old-school arbitrageurs, Friedmanites and central bankers, for example—see money as the primary asset or value. The arguments put forward in the previous paper might be more persuasive to such folk if we were to convert the opportunity cost of time into monetary terms. So let's put a price on time, in particular, on the time that verbosity and obfuscation costs industry and commerce.

Benjamin Franklin, one of the Founding Fathers of the United States, is credited with the well-worn adage that time is money:

> "Remember, that *time is money*. He that can earn ten shillings a day by his labor, and goes abroad, or sits idle, one half of that day, though he spends but six pence during his diversion or idleness, ought not to reckon that the only expense; he has really spent, or rather thrown away, five shillings besides."[1]

So what money does society throw away in tolerating the communicative imperfections (such as obfuscation and verbosity) that it does? Most workers—whatever the shade of their collar—spend some part of their working day engaged in reading: reading reports, emails, operating instructions, requests for tender, policies and so on. For some it might be little more than 20 minutes a day; for others it might be five or six hours. Suppose that, on average, 15 minutes of that reading time is spent disentangling the intended meaning from poorly expressed language (including the time taken to send emails or

---

1. From a letter to an unnamed tradesman, dated 1748. Emphasis added.

Adapted from a training workbook written to accompany technical writing workshops conducted by Abelard Consulting. Written between 2009–15. Addendum adapted from Marnell 2015, chapter 5.

make phone calls seeking clarification) and reading words that are unnecessary. Since those 15 minutes are on top of the time they would have taken to read everything had everything been immediately obvious, we can put a figure on the opportunity cost of poor writing: 15 minutes per day. That is unproductive time, time that could have been put to more profitable uses. In countries where the typical working week is 40 hours, this equates to approximately 3% of paid time. Not much, perhaps—until you see the figure converted to currency and applied nationwide.

Take Australia as an example. As at May 2010, there were approximately 11.5 million employees of which 63.3% (or 7.2 million) were in full-time employment. The average weekly earnings of full-time employees was $1313 (that is, $68,000 per annum). Three per cent of $68,000 is $2400. Thus the opportunity cost of poor writing in Australia—the money that could be put to uses other than teasing out the meaning of impenetrable texts and wading through verbosity—is at least $14.75 billion per year: $14,750,000,000. And that is not counting part-time employees.

Based on a comparison of populations alone, the opportunity cost of poor writing in the United Kingdom is in the order of £25 billion, in Canada in the order of CA$23 billion, and in the USA in the order of US$210 billion. That is, in total, close to US$300 billion every year in lost productivity. *That figure is huge.*[2] Surely a billion or two a year in remedial English classes wouldn't go astray.

## Addendum: Thrift in the currency of words

We might have written a clear text—one free of nonsense, ambiguity and vagueness—and have used language that is familiar to our intended audience, but still our attempt at communicating might fail. It might fail if we overlook the

---

2. About 5600 new high schools, or 10,000 middle schools, could be built each year for this money. (Based on US construction costs reported at http://www.ncef.org/ds/statistics.cfm#.)

*context* in which the audience might read our text. For that context could include aspects that will compete for the attention of our readers, aspects such as *demands* and *distractions*. Consider firstly competing demands. Despite the immense labour-saving potential of computers, most of us are time-poor. The leisure dividend most of us hoped for never materialised. It was converted instead into profits. Moreover, we appear to be working more than we did before the advent of the computer (which will strike many as paradoxical). No-one, it seems, has an empty inbox, work is taken home on the weekends, recreational leave is delayed and personal stress is on the rise. This is the unfortunate backdrop against which we must frame our attempts to communicate. The more words there are to read, the longer it takes to read them. That is a truism. In a climate of time-pressure, a reader who detects too much verbiage might become alienated from the text and disengage from it. They sense that it is needlessly, discourteously keeping them from other pressing demands. They might abandon their reading (even if they would be better off if they had persevered). In a word, communication breakdown has occurred—the antithesis of the very goal that prompted the writing in the first place. Thus the wise writer assumes that their audience is demand-stressed and time-poor, and chooses to write with thrift in mind, that is, economically. For by doing so—by avoiding verbosity, redundancy, nominalisation and other forms of bloat—they increase their chances of being read.

It is true that sometimes we don't want an author to be economical with words. The author of a novel, for example, might be such a brilliant anatomist of character, setting and plot that we don't want the book to end. Likewise, a poem may be so luscious of word-music that we want more, not less:

> "I have seen them riding seaward on the waves
> Combing the white hair of the waves blown back
> When the wind blows the water white and black."[3]

---

3. From *The Love Song of J. Alfred Prufrock* by T. S. Eliot, 1915.

"It is spring, moonless night in the small town, starless and bible-black, the cobblestreets silent and the hunched, court-ers'-and-rabbits' wood limping invisible down to the sloe-black, slow, black, crowblack, fishingboat-bobbing sea."[4]

But if what you are writing is informational rather than fictional (as is technical writing) and you suspect—as you must—that your intended readership is time-poor, writing *economically* becomes a practical necessity. If you want to communicate but deliberately take 20 pages to convey what could have been conveyed in 15, your readers may not have the time to read all that you have written (including the more useful or important information they were hoping to find). In that case, you have engineered communication breakdown. Your writing was self-defeating. Other demands on your readers' time have out-competed your wish to communicate with them. That may not be entirely their fault; some of it must lie with you.

As well as competing demands there are competing *distractions*. The magnetism of cheap gadgets capable of doing the seemingly impossible, the loneliness-breaking offerings of social media, the ready access to global television and other awe-inspiring inventions combine to tempt readers away from reading (or at least from careful, unbroken reading). Writers need to accept that their texts will never out-compete the dazzle and lure of technological wizardry. Thus their best approach, if they want their works to be read in full, is to adopt the philosophy of minimalism, that is, writing economically. It is simply a practical necessity in this time-demanding, gadget-distracting 24/7 world we find ourselves in.

Writing economically might also be a *moral* necessity. We all have a finite number of hours in a lifetime, and we all want time away from study and from the office. We want—and need—time to engage in activities that give our life special richness and significance, whatever they might be: time with our friends and family, time to hike through the bush, time to

---

4.   From *Under Milkwood* by Dylan Thomas, 1954.

visit galleries, and so on. I suspect that on reflection most people would assess time as the asset they value over all others. (If the word *asset* carries too much of an economics overtone for your liking, then think of time as your most valuable *possession*, a possession that is a birthright.) Most of us have other assets, assets more tangible than time, but of less intrinsic value. We might have a classic car, a collection of first-edition books, a plasma television, expensive jewellery, and so on, and we would feel aggrieved if some such asset was stolen or damaged. (We might have some of these assets insured so as to lessen the sense of loss in the event of theft or damage.) But just how valuable would that car or book collection be if we had no time to drive or read. The value of all possessions presupposes time, thus elevating time to the most valuable possession of all.

Now the theft of a tangible asset—such as a car or a painting—is a moral issue. Who can deny that? But if so, then so must be the theft of an even more valuable asset: time.[5] Morality is largely concerned with the prevention of harm to others: physical harm, mental harm or being made worse off. By having time taken from us, we are made worse off. We now have less time to do the things that are most valuable to us. It is a moral issue. Thus readers have the right to feel aggrieved at having to plough through large tracts of scarcely penetrable verbiage in order to get the information they need. They are having their time—a finite resource of great value—stolen from them, time they could be putting to better use.

Let's clothe this idea with some detail. Suppose you have written a 50-page report and every sentence in it has three superfluous words (a feat that is easily achievable). If we assume that the average length of a sentence is 18 words (James 2007, p. 354), your report effectively contains just 42 pages of substance. That is, the accumulated verbiage—all those unnecessary words—accounts for about eight pages. A reader

---

5.  Time might be an *intangible* asset, but that doesn't necessarily rule it out of the moral sphere. There are other intangible assets that owners have a moral right to, such as copyright, patents and goodwill.

might take 40–45 minutes to plough through that verbiage—much more than the conservative 15 minutes we used in our earlier calculations—time they would not have needed to devote to the report had it been written with concision in mind. Thus you have stolen time from every one of your readers. (It doesn't matter that readers might not notice that their time is being stolen. It is still a moral issue, just as taking money from the wallet of a blind person is still a moral issue even if they will never know that money has been taken, and even if they have more money than they need.)

Thus there are moral as well as practical reasons why economical writing is essential. In the ever-demanding, non-stop, multi-channel world in which most of us live today, the demands on our time are often excessive. If we want to increase our chances of being read and getting a message across, we should assume that our intended audience is time-poor and that our communication is likely to be one in a pool of many communications each competing for the attention of our audience. Moreover, even if our readers have time to read what we have written, it is immoral to assume that we can take up as much of their time as we like.

This brings us to a very important principle of good writing:

> *Principle of economy*: to maximise your chances of being read, avoid verbiage and other forms of bloat that will steal time from your intended readers.

Let's finish by illustrating some common types of verbiage: verbosity, triviality, redundancy, tautology and nominalisation.

*Verbosity* occurs whenever a string of words is used when a single word (or a shorter string) would suffice. Some examples:

> You write *at this point in time* instead of *now*, *has the ability to* instead of *can* and *owing to the fact that* instead of *because*.

*Triviality* occurs when the writing states the blindingly obvious, as in:

> Connect the unit to mains power. Use a wall socket that is accessible. [As if it was possible to use a socket you couldn't access!]

*Redundancy* occurs when you use words to do work that is already being done by other words in the sentence, as in:

*There are four different flavours.* [If they weren't different, you would have one flavour, not four.]

*Tautology* is a type of redundancy where a word repeats the meaning of another word used in the sentence. The writer is in effect repeating themselves, and consequently wasting readers' time. For example:

Limbitin is to be taken orally through the mouth. [So what does *orally* mean?]

*Nominalisation* is turning a verb into a noun or noun phrase and replacing it with an irrelevant verb (usually some grey, all-purpose verb that has nothing to do with the action that is being expressed). For example, you want to write:

We studied the effect of DDT on brassica vegetables.

but find yourself writing instead:

We undertook a study of the effect of DDT on brassica vegetables.

A relevant verb (*studied*) is converted into a noun phrase (*a study of*) and replaced by a verb unrelated to study (*undertook*) Nominalisation always introduces unnecessary words and, if not reined in, will deaden most readers' enthusiasm for reading. It is one of the forms of writing now forbidden to writers of public service documents in the United States.[6]

Let's conclude by repeating two important cost-related facts:

- An employee spending just fifteen minutes a day ploughing through or deciphering verbiage contributes a 3% productivity drain, a drain that can easily accumulate into a hidden impost of millions of dollars a year for an organisation.
- Just three superfluous words in every sentence in a any document generates about 17% of waste. Such extravagance in a document of 50 pages equates to 8 superfluous pages. Many of us would need 2–3 hours to type 8 pages.

---

6.  See *Plain Writing Act 2010*, signed into law on 13 October 2010.

In an era of brimming inboxes, who has a spare 2–3 hours to do that? And why should readers have to submit to an unnecessary drain on their finite time as they read that verbosity?

Time-thievery is like a communicable disease. It spreads from writer to reader. The writer allows their own time to be robbed with unnecessary verbosity, and subsequently readers have their own time robbed by having to plough through that verbosity. The cure is, clearly, *word hygiene*.

# 4: Does sentence length matter?

Those who come to technical writing from a scientific or technical background are often puzzled by the ready acceptance of guidelines that have never been subjected to critical scrutiny. They read that the active voice is always to be preferred, that paragraphs must be kept short, that numbers must be spelt out if less than ten, and so on—and they search in vain for the research that might give these guidelines the imprimatur of scientific rigour. They find plenty of language pundits eager to share their beliefs, but far fewer experts offering sound reasons.

One set of guidelines crying out for scrutiny has to do with sentence length. Many a pundit is convinced that the number of words in a sentence should fall within a particular range. Some also contend that sentences should not have more than a specified number of words. For example, Dr Anetta Cheek tells us that we should "aim for an average sentence length of between about 15 and 22 words". Moreover, we should "avoid sentences of more than 40 words" (Cheek 2010, p. 10). Dr Neil James, Director of Plain English Foundation, is also convinced that average sentential word-count is an important consideration in good writing: "a 15–20 word average is *fundamental* to writing well" (James 2007, p. 244. Emphasis added). Further, James hints that 35 words is probably the maximum we should include in a sentence: "if your content is complex, or if you need to include a subsidiary point, you will

First published in *Southern Communicator*, issue 33, October 2014.

want to go over 18 words — perhaps even stretching towards 35 words".

The view that sentential word-count is an important consideration in writing well does have prima facie plausibility. We do get lost in long sentences (and probably no-one has ever fully understood that famous 823-word sentence in Victor Hugo's *Les Misérables*). We attempt to make sense of a sentence with the mechanisms of our short-term (or working) memory. That memory is limited by our biology. If a sentence we are reading surpasses that limit, ideas we read earlier in it are pushed out of memory and we forget part of what we have just read. We will need to process the sentence in parts and this will require rereading (perhaps a number of times). The respectful writer — keen to communicate while imposing the least effort on readers — will try to avoid making readers reread, and they will do so by ensuring that the sentences they write do not exceed the capacity of our short-term memory.

The common view is that we do this by limiting our word count. Yet experiments that challenge this view are simple to construct. In my writing classes over the last few years I have displayed on a screen a simple 11-word sentence. The sentence remained visible for six seconds (more than enough time for the sentence to be read by any competent reader). Students were asked to read the sentence just once and then to write down its main points. They were told not to fuss about reproducing the sentence exactly as it was presented, but just to jot down its gist. I then displayed another 11-word sentence for six seconds. Again students were asked to jot down the main points. Here are the two sentences:

[A]    The United States of America has reluctantly signed the peace accord.

[B]    The hot, treeless plain is covered in dark, smooth, elongated rocks.

I then gathered the cards the students had written on and tallied the number of points that had been correctly recalled. Despite the sentences being of identical word-count and of comparable vocabulary, average recall scores — that is, the average number of correct responses given—were markedly different: sentence [A] a little over 89%; sentence [B] just 33.9% (from 519 students). If word count is a critical factor in reader comprehension, then surely these sentences should have yielded similar recall scores.[1] The explanation for the discrepancy is simple, and it does have to do with the limits of our working memory.

The notion of a limit to working memory was made popular in a 1956 literature review by the American psychologist George Miller. Miller thought the limit, averaged over diverse stimuli (such as words, numbers, sounds, etc.), was *seven* distinct chunks. Subsequent research indicates that the average is closer to four (a fact that torpedoes a major claim of the Information Mapping fraternity):

> "the capacity of short-term memory ... is known to be quite small, only about four chunks." (Kintsch & Rawson 2005, p. 224)

> "a central working memory faculty is limited to 3 to 5 chunks for adults." (Cowan 2010, p. 52)

To psychologists, a *chunk* in the act of verbal comprehension is a basic unit of meaning. Moreover:

> "the basic units of meaning are propositions. Propositions are n-tuples of *word concepts*, one of which serves as a *predicator*, and the remaining ones as *arguments*, each filling a unique semantic role. The predicator specifies a relationship among the arguments of a proposition. For instance, in the proposition (LOVE, Experiencer: GREEK, Object: ART) there are two arguments, GREEK and ART, and the predicator LOVE; in English this proposition could be realised with the sentence *The Greeks loved art*. It is important to note that the arguments of a proposition are *concepts* rather than words." (Kintsch et al. 1975, p. 196)

---

1.  Note too that the sentences used in the experiment have a word count well below what many language pundits—Cheek, James and others—consider the upper limit for immediate comprehension.

In this example, the basic unit of information has the S–V–O structure, that is, subject–verb–object: The Greeks (S) loved (V) art (O). But there are other equally basic structures. For example, S–V (as in "Emily laughed"), A–N ("bacterial infection") and A–V ("loudly abused").

What cognitive science tells us, then, is that if we want our readers to understand our sentences on one reading, we need to limit the basic units of information in them to no more than four. And this explains the result of the experiment I described above. To see how, dissect each sentence into its constituent chunks:

[A]     The United States of America has reluctantly signed the peace accord.
        Two chunks: <USP> and <SR> — (a) the United States of America has signed the peace accord and (b) the signing was done reluctantly.

[B]     The hot, treeless plain is covered in dark, smooth, elongated rocks.
        Six chunks: <P, C, R>, <P, H>, <P, T>, <R, D>, <R, S>, <R, E> — (a) the plain is covered in rocks, (b) the plain is hot, (c) the plain is treeless, (d) the rocks are dark, (e) the rocks are smooth and (f) the rocks are elongated.

The variance in the rates of recall—89% for [A] and 33.9% for [B]—is due to the fact that the sentences, though of identical word count, have significantly different chunk counts. Sentence [A] poses little cognitive strain on our working memory because it has just two chunks, well within the four-chunk limit of our working memory. On the other hand, sentence [B] forces the reader to attempt to squeeze six chunks into a four-chunk slot and, as a result, some chunks escape. In the experiment, 78% of students correctly recorded both chunks of information in sentence [A] whereas only 1.9% correctly recorded all six chunks in sentence [B].

A fixation on word-count—average or maximum—is thus misleading. I could write a document where the average sentence-length is between 15 and 20 words—a so-called "fundamental" average—no sentence is longer than 35 words, the vocabulary is familiar (as is the grammar and punctuation) and yet every reader will have to read every sentence at least twice to fully understand it. All I need do is pack more than four chunks of information into every sentence. My writing meets the pundits' guidelines, but it could hardly be considered good writing.

So length does matter—*so long as we are measuring the right thing*. And the right thing is not words, but chunks: the basic units of information that one is trying to impart to readers.

As we've seen, short sentences can be unreasonably burdensome on readers; conversely long sentences—longer than the limit recommended by many pundits—can be a breeze, so long as the chunk limit is kept to four or less. Here is an example:

> The Democratic People's Republic of Korea is smaller than the People's Democratic Republic of Lao, the People's Democratic Republic of Lao is smaller than the Republic of South Africa, the Republic of South Africa is smaller than the Democratic Republic of the Congo and the Democratic Republic of the Congo is smaller than the United States of America.

The logical structure of this 58-word sentence is very simple: A < B, B < C, C < D and D < E, and its apparent wordiness does not impede its immediate comprehension. This is because the concepts are simple—mere countries—despite their names being multi-word compound nouns. It is pertinent here to repeat the last sentence in the quote above from Kintsch:

> "It is important to note that the arguments of a proposition [that is, a basic unit of information] are *concepts* rather than words."

Finally, consider the readability scores in Microsoft Word (scores that have been used by courts of law in the USA to justify awarding damages against writers). The Flesch reading ease score—and its derivative, the Flesch–Kincaid Grade Level—are based solely on the average number of words per sentence and the average number of syllables per word in a given text. In most cases, the reading ease score will fall between 0 and 100: the higher the value, the more readable the text, or so the theory goes. So what Flesch scores do our two eleven-word sentences get? Sentence [A], with just two chunks of information, gets a score of 41.8 while sentence [B], with six chunks, gets a score of 72.6. So the sentence that the vast majority of readers will have to read twice before they fully understand it is deemed more readable than the sentence that nearly everyone understands immediately on first reading! Proof yet again that the readability formulas in Microsoft Word are best thrown onto the scrap-heap of pseudoscience.[2]

As the technical writing profession lurches towards accreditation, it is vital that the knowledge on which writers will be judged is evidence-based, that is, stamped with the imprimatur of scientific rigour. It is vital too that those who contribute to standards on technical writing ensure that their recommendations shine with the burnish of proof. It will backfire on the profession if later we find that our cherished guidelines were at best hunches, old wives' tales or stylistic prejudices.

---

2. The Flesch reading ease score is discussed in detail in chapter 6, "Can the quality of writing be measured?" starting on page 75.

# 5: Pitfalls in procedure writing

Procedure writing is the bread-and-butter of technical writing. Most technical writers spend most of their working time writing procedures: instructions to explain how to use or service a product. Writing procedures is not rocket science. In large measure, it is just the application of primitive logic: what steps need someone take to achieve such-and-such a goal?

But if we want to write procedures from our customary *audience-centric* perspective—the perspective that in large part defines our profession—then we need to consider more than just *any* way of achieving some goal and letting that pass as a worthy procedure. We need to think about, and respect, our readers' safety and their time. We need to look out for our readers, warning them of risks and preventing them from rushing into unwanted circumstances. And we need to minimise the effort—physical and mental—our readers need to achieve their practical goals. Might it be easier, or quicker overall, to add in another step that, although logically unnecessary, is practically beneficial (such as a step to remove some unrelated component to make it easier for the user to get at the component they need to get at)? Or might it be easier or quicker for the majority of my likely readers to put one particular step before another?

These are issues that go beyond mere language. But they are issues that, when considered and rigorously applied, can

First published in *Words*, vol. 2, iss. 3, October 2010

turn an everyday workable procedure into a gem of simplicity and efficiency.

Our readers will rarely notice the extra effort that has gone into a well-considered procedure, and thus our extra efforts will be unappreciated. But that is also the case with writing in general. Clear, transparent writing mostly goes unnoticed. (When they read a well-written sentence in a novel or a user guide, readers rarely appreciate the behind-the-scene labours of drafting and redrafting, watered by the sweat of writer's block and editor surveillance). A procedure is much like a sentence in this regard: its structure is only noticed if there is something odd, unexpected or careless about it.

In what follows are some common problems in procedure writing, problems that get in the way of our overarching goal: to have our readers complete their tasks with the least effort and least risk. The list of problems is in no way complete, and no list might ever be complete. But the point of this article is not to provide comprehensive advice on procedure writing. Rather it is to underscore the sometimes overlooked fact that logic as much as language is needed in writing audience-centric procedures.

## Placing the condition after a conditional action

A conditional action is one that needs to be done only if a particular condition is the case.

6. If you want to delete all the records, press **OK**.

6. Press **OK** if you want to delete all the records.

In this example—one direction expressed in two ways—the action of pressing **OK** is *conditional upon* the user wanting to delete all the records. But note how very different—practically as well as linguistically—the two versions are.

In the first version, the conditional clause (the *if* clause) precedes the imperative (or action statement); in the second it follows the imperative. That is the linguistic difference. But consider the *practical* difference, namely, the difference in the probabilities of readers doing something they didn't intend to do. The hurried user—as most of us are these days—is more likely to unintentionally do the action if the step is expressed in the second way—with the imperative before the condition—than in the first. They see the command and, swept along by the maddening rush of today's Zeitgeist, obey it in the split-second between reading it and reading the condition that follows, that is, before realising that they actually had a choice, and that their choice might well have been *not* to do the action. (Oh damn! I didn't really want to delete all the records.)

We should never assume that all our readers will have the time for slow and careful reading, free of breath on the neck. Thus in procedure writing, a conditional clause should always be placed *before* the imperative: "if $x$ is the case, do $y$" and never "Do $y$ if $x$ is the case".

## More than the minimum number of steps or clauses

Economy is a worthy goal and not just in our choice of words: it is also a worthy goal in designing procedures. The least number of steps—or clauses—a user has to read in order to complete the procedure the better, *ceteris paribus*.

### Example 1

Can you see the problem with this snippet?

  5. If you have Windows 98 installed, do …

  6. If you have any other version of Windows installed, do …

The problem is this: almost no one these days has Windows 98 installed, so almost every reader will have to read a step (or a clause in a step) that is of no relevance to them. And that is wasting their time. It is far better to reverse the order of these two steps (and tweak the text a little). Now the majority of readers will know what to do after reading just one step.

## Example 2

What is the problem with this similarly flawed snippet, aimed at a general readership?

1. If you are male and under 35, do $X$ and go to step 4.
2. If you are male and 35 or over, do $Y$ and go to step 17.
3. If you are female, do $Z$ and go to step 28.

One way to approach audience-segmented procedures like this is to employ what might be called the *effort index*. The effort index is the percentage of your overall audience multiplied by the number of clauses that that segment has to read before they learn what to do. So, as currently structured, youngish males have to read two clauses. They constitute about 25% of my audience (assuming a three-score-and-ten lifespan) and thus their effort index is $25 \times 2 = 50$.[1] The rest of the males have to read four clauses, so their effort index is 100. The females — being 50% of my readership — will have to read three clauses before they know what to do and thus their effort index is 150.[2] So the total effort index for the current structure is 300.

But notice that we can reduce the overall effort index by rearranging the steps:

1. If you are female, do $Z$ and go to step 28.
2. If you are male and under 35, do $X$ and go to step 4.
3. If you are male and 35 or over, do $Y$ and go to step 17.

Now the effort indices are 50 (female), 75 (youngish males) and 125 (older males), yielding a total of 250.[3]

---

1. The two clauses are "If you are a male" and "[if you are] under 35".
2. The three clauses are "If you are a male", "If you are a male" and "If you are a female". We have approximated the true proportion of males and females in the world so as to simplify our calculations without sacrificing informativeness.
3. Of course, steps 2 and 3 could be reversed without affecting the result (but only on the assumption that males under 35 are just as numerous as males 35 and over).

But we have not finished yet. We can reduce the effort index further by deleting a step and recasting our snippet as:

1. If you are female, do Z and go to step 27.
2. If you are male and:
   - under 35, do X and go to step 3.
   - 35 or over, do Y and go to step 16.

The effort index is now 225.[4] Indeed, by adopting this structure—where one clause is suppressed but still implicit—it does not matter whether the females or males are placed first:

1. If you are male and:
   - under 35, do X and go to step 3.
   - 35 or over, do Y and go to step 16.
2. If you are female, do Z and go to step 27.

You could go even further. Consider the following rearrangement:

1. If you are female, do Z and go to step 28.
2. If you are under 35, do X and go to step 4.
3. Do Y and go to step 17.

The effort index is now just 150. This version is vastly more economical, with an effort index half of what we started with. Part of the simplification is due to the fact that we have managed to remove all the "If you are a male" clauses. Including them at the start of steps 2 and 3 would only be stating the blindingly obvious, because only males will read steps 2 and 3.

The effort index is a useful measure of how economical the structure of a procedure is. If technical writers are concerned—as they should be—with minimising their readers' cognitive load and time leakage, this index should be in all our documentation development tool-kits. Still, if you suspect that your audience would find a particular structure confusing despite its economy—as some might, perhaps, our bare-bones rendition considered in the previous paragraph—then it's

---

4.  $(50 \times 1) + (25 \times 3) + (25 \times 4) = 225$.

incumbent on you to add some meat to the bones, turning, for example, implicit antecedents into explicit conditions where necessary. As in all aspects of technical writing, clarity always trumps economy.

## Example 3

Imagine two commercial greenhouses (A and B) in each of which grow various vegetables. Twenty degrees Celsius is the optimal temperature for growth in both A and B. In fact, the temperature should never be allowed to fall below 20, although sometimes it does. And sometimes it rises above 20 (which, although not optimal, is better than the temperature falling below 20). And often the temperature in A is different to the temperature in B. Moreover, temperatures are just as likely to be lower than optimal as they are to be higher than optimal.

Twice a day gardeners observe the temperature in A and B and adjust it if necessary. But they have just a single thermostat at their disposal, and whatever change of temperature it brings about in A, the same change occurs in B.

Your task is to write a procedure to instruct the gardeners what to do each time they check the temperature in each greenhouse. How do you proceed?

One obvious way is to list all the possibilities and write a procedural step for each one. In our example, admittedly contrived but still instructive, there are nine possibilities:

- A > 20 and B > 20, = 20 or < 20
- A = 20 and B > 20, = 20 or < 20
- A < 20 and B > 20, = 20 or < 20

Hence our procedure could take this exhaustive[5] form:

1. If A and B are both greater than 20, reduce the temperature until the temperature in the cooler greenhouse (or both) is 20 degrees.

---

5.  Pun intended.

2. If A is greater than 20 and B is equal to 20, leave the settings as they are.

   ...

9. If A and B are both less than 20, increase the temperature until the temperature in the cooler greenhouse (or both) is 20 degrees.

But this is fairly tedious: more so for the reader than the writer. For most conditions, four or more steps have to be read before the gardeners know what to do. But there is a better way.

If you consider which of the nine conditions *requires the same action*, it is possible to reduce this procedure to just three steps:

1. If A and B are both greater than 20, reduce the temperature until the temperature in the coolest greenhouse (or both) is 20 degrees.

2. If either A or B is less than 20, increase the temperature until the temperature in the coolest greenhouse (or both) is 20 degrees.

3. In all other cases, leave the settings as they are.

All conditions are covered: step 1 covers one condition; step 2 covers five and step 3 covers three. But now the most number of steps anyone needs to read in order to discover what to do is three: a vast improvement.

Here is a way of visualising the possibilities using a decision table:

|   |     | B |   |   |
|---|-----|-----|-----|-----|
|   |     | =20 | >20 | <20 |
| A | =20 | x | x | y |
|   | >20 | x | z | y |
|   | <20 | y | y | y |

where $x$ = leave the settings as they are, $y$ = increase the temperature until the coolest greenhouse (or both) is 20, and $z$ = reduce the temperature until the coolest greenhouse (or both) is 20.

The fact that there are only three discrete actions ever needed should alert you to the strong possibility that only three steps will be needed in the procedure.

## Being overly economical

In the last section we considered the importance of economy in procedure writing. But economy—understood as minimising the number of clauses or steps needed for the majority of the anticipated audience to discover what to do—might not always be what would please the majority of that audience.

### Example 1

If economy of steps was our overriding goal—rather than minimising our readers' effort—we might be tempted, in very complex scenarios, to concatenate many conditions into the conditional clause introducing a step:

3. If A is true and either B or C is true or B is false but D is true, do … and go to step 10.

Just as a sentence starts to become indigestible after the third clause (and sometimes before), a multi-condition conditional clause in a procedural step (as above) might well over-tax the cognitive abilities of many of your readers. Keeping a sentence to no more than two clauses is sound advice (overlooked by most academics). The parallel advice— keep your conditional clauses to no more than two conditions—is equally sound advice. There is no point trumpeting the economy of your procedure if most of your readers cannot fathom its conditional complexities.

### Example 2

Suppose, for instance, that you are writing a manual that explains how to service or replace certain parts in a piece of equipment. An economical procedure might state:

5. Remove the trim covering the power supply recess.
6. Remove the screws fastening the cable to the rear of the power supply.

But if only those with small hands and fingers are likely to reach with ease the screws at the rear of the power supply, then your economy of steps might have been purchased at the cost of time: the time it takes for the average service engineer to get those screws out. It may well have been easier for the majority of readers if you had added another step:

5. Remove the trim covering the power supply recess.

6. Remove the bracket securing the power supply to the main casing.

7. Slide the power supply forward a little and remove the screws fastening the cable to the rear of the power supply.

There is an extra step here, but it might save the majority of readers time, not to mention the pain of excoriated knuckles as they attempt what turns out for them to be next to impossible.

So economy is not just about minimising the number of steps in a procedure. That goal, laudable in the abstract, is merely a qualifiable component of the greater goal of minimising our readers' time and effort.

## Undifferentiated literals

A literal is a word or string of words that the reader will see on the screen or on the product being documented. A field name is a literal, as is an error or confirmation message. If a step in your procedure must refer to a literal (as will often be the case), help the reader locate the literal on the screen or product by applying some typographical cueing. For example, you might adopt the convention—quiet common in contemporary technical writing—of setting field names in bold text.

Without some form of typographical cueing, the literal will appear undifferentiated from the surrounding text, and this could waste readers' time as they determine what exactly the instruction is. Consider the following example:

4. Select GST and FBT and press Enter.

4. Select **GST and FBT** and press ENTER.

The second example makes it immediately clear that there is just one option to select, not two (as the first example might suggest).

Typographical cueing is also commonly applied to the names of keys (as in the example immediately above) and to text that the user needs to enter.

> 7. Before the pressure reaches 100 kPa, type parameter 1887 and press **Enter**.

> 7. Before the pressure reaches 100 kPa, type `parameter 1887` and press ENTER.

Without typographical cueing, what is the reader meant to type in: 1887 or parameter 1887?

## Branches not explicitly terminated

A procedure may have discrete branches or paths. At step 5, say, the user might be able to choose one of two paths. Steps 6, 7 and 8 take them down one path; steps 9, 10, 11 and 12 take them down another. The step might look like this:

> 5. If you are running version 10 or earlier of [some software], do steps 6, 7 and 8; otherwise do steps 9, 10, 11 and 12.

You should make it crystal clear after step 8 that the reader who is using an early version of the software has completed the task and can ignore the rest of the procedure. (In other words, there must be no chance that such a reader will think that step 9 is necessary once they have finished step 8.)

Here is one of a number of techniques to achieve this:

> 8. Click OK.
>
> You have now finished the pay-run. *Ignore the rest of this procedure.*
>
> 9. Select employees who have resigned in the last month.

The indented statement below step 8 could be called a *buffer statement* (as it is analogous to the buffer stop at the end of a railway line that prevents trains from derailing).

If you think that adding a buffer statement is over-egging the pudding, imagine yourself in a busy office. You have just read step 5 and now know that you only have to do steps 6, 7 and 8. You are about to commence step 6 when the telephone rings. Twenty minutes later you return to the procedure, finish step 6 and start step 7 just as your manager wanders in for a report on your project's progress. Thirty minutes later you return to the procedure, finish step 7, take a desperately needed powder-room break and return to your desk and tackle step 8. You finish it and start step 9 and wonder why there is a mismatch between what is on your screen and what is described in the procedure. Understandably, you didn't remember that step 8 is as far as you needed to go. Some would, but many wouldn't. And that's why the 15-seconds-to-write buffer statement in an embedded branch of a procedure is a courtesy many readers will appreciate.

## Late prerequisites

Do not delay specifying a prerequisite until the step that requires the prerequisite. All prerequisites should be specified in the preamble to the procedure so that the reader can attend to the prerequisites before commencing the procedure.

> 13. Select **GST liability**. This is possible only if you have clicked the **Business** option in the **Preferences** window — see 'Setting tax options' on page 23.

With a step like this, some users will have to back out of the procedure and start all over again once the prerequisite has been attended to, that is, once they have changed the specified preference. This is wasting the user's time.

As a general rule, you need to tell the reader *in the preamble to a procedure* what tools, spare parts and information they need, and what previous actions they should have done. No engineer will be impressed with your bogey-removal procedure if you tell them, at step 20, to use a rattle gun to loosen the nuts on the bogey if it means that they need extricate themselves from the maintenance well, clean up and traipse back to the tools storeroom to find a rattle gun. You are just wasting their time.

## Poorly placed risk messages

A topic akin to late prerequisites is poorly placed risk messages: but it is much more serious. Risk messages are commonly placed:

- in the preamble, if the risk is especially serious, and
- before or with the step at which they apply.

Placing a serious risk message in the preamble can help prepare the reader for the task ahead. Knowing that there are serious risks *before* they embark on a procedure is likely to cause the reader to approach the task with the right frame of mind — with seriousness, concern and care — which may be unlikely at 4.30 on a Friday afternoon.

Repeating the risk message at the step where there is a risk reinforces the possibility of unwanted results. Moreover, such messages will be the only notification of possible risk for those hurried readers who skip the preamble.

Where the action itself is risky (as opposed to the state that the action brings about), it is imperative that the risk statement *precedes* the action. The following snippet — which breaks this rule — shows no concern for the welfare of the reader:

12. Turn the hydraulic key anticlockwise. The jack, and car, return to ground position.

    **WARNING**: Make sure that you are not underneath the car when you perform this step.

Too late, she cried.

# 6:  Can the quality of writing be measured?

Science and measurement are inseparable. Without measurement, science would not exist; with measurement, our pattern-sensitive minds convert apparent randomness into scientific hypotheses. *Homo sapiens* love to measure. We measure time—occasionally obsessively as a weekend draws near. We measure our weight, our girth and our cholesterol level, and fret over the values. We measure our wealth, the economy of our cars, our electricity usage. To another species, we might seem that we have a quantification fetish. If it can be measured, *Homo sapiens* measure it. Some things can't be measured, as Heisenberg made clear with his uncertainty principle: the velocity and momentum of an electron, for example. There are also some things we can't measure not because of inherent quantum uncertainty, but because of logical boundaries. (Can we measure the width of the universe? How would we know when we have reached an edge of the universe?). But these limitations haven't dampened our passion for quantification.

But measurement is not always straightforward. Our instruments might lack the desired sensitivity, our tools for analysing the data might be irreparably blunt (mere statistics, for instance) or our measurement might need to be made indirectly (thus multiplying the risk of imprecision). But a greater problem is where it is difficult to pin down what we want to measure. What would you look for if you wanted to measure happiness? (Wealth? Friends?) Or the wealth of a

First published, in two parts, in *Southern Communicator*, issues 14–15, 2008 under the title *Measuring Readability*. Additional material adapted from Marnell 2015.

nation? (GDP alone? GDP less the cost of externalities? Social capital as well as material wealth?) Sometimes the issue is purely academic. Nothing much hangs on how the concept is defined (so long as it is defined). But when the outcome of our fetish for quantification is put to uses that hinder, harm or deny, it is paramount that we know, and agree on, what the concept is that we are measuring. One such concept is intelligence (or IQ). Society seems addicted to measuring IQ and distributing society's benefits—such as scholarships, jobs and memberships to certain closed societies (such as Mensa) on the basis of one's score on an IQ test. Sadly, IQ tests have also been used to justify eugenics. (Gould 1996)

Another concept that has been subjected to dubious quantification is *readability*. Readability, like usability, is one of the central themes in the quest for good writing. To maximise readability is a goal that every writer, technical or otherwise, should strive to achieve. To argue otherwise is tantamount to arguing that we do not write to communicate, a bizarre view to say the least.

Most of us have at least a vague idea of what readability is. When we look at an act of parliament, or legal contract, drawn up before Plain English became popular we wonder how anyone could have deciphered it. However difficult a current act of parliament or legal contract might be, the difficulty pales into insignificance when compared to the difficulty inherent in like documents of yesteryear.

There have been attempts to provide a way of measuring readability. Some word processing software will give you a readability score (being an assessment of how easy it is for readers to read the document that is currently on the screen). An example is shown in figure 6.1, taken from a recent version of Microsoft Word.[1]

---

1.  To see your scores for a particular document, you will need to have selected **Show readability statistics** on the **Proofing** tab of the **Options** window and then run a grammar check of the document.

Readability Statistics

Counts
Words                           141
Characters                      779
Paragraphs                       17
Sentences                        14

Averages
Sentences per Paragraph         1.2
Words per Sentence              8.7
Characters per Word             5.0

Readability
Passive Sentences               7%
Flesch Reading Ease            30.7
Flesch-Kincaid Grade Level     11.1

OK

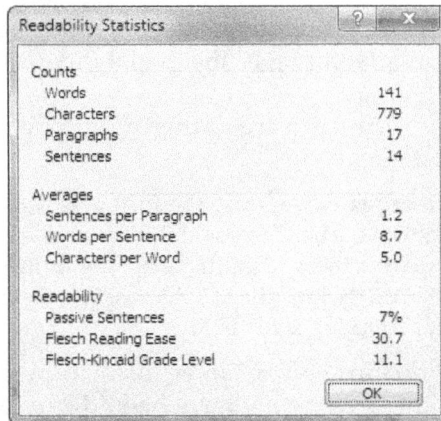

Figure 6.1  Sample readability statistics in Microsoft Word

*Readability* has two general senses, one applying to document design, the other to language. Readability as it is applied to document design is concerned with such matters as line length, leading, white space, font type and the like. Readability as it is applied to language is concerned with *comprehensibility* or *understandability*:

> "*Readability* means *understandability*. The more readable a document is, the more easily it can be understood…" (Samson 1993, p.58)

Another quote:

> "[Readability is] the efficiency with which a text can be comprehended by a reader, as measured by reading time, amount recalled, questions answered, or some other quantifiable measure of a reader's ability to process a text…" (Selzer 1983, p. 73)

And one more definition of *readability*:

> "[the] sum total (including all the interactions) of all those elements within a given piece of printed material that affect the success a group of readers have with it. The success is the extent to which they understand it, read it an optimal speed, and find it interesting."[2]

In this chapter we are focusing on readability as it pertains to language, not as it pertains to document design. And this is the variety of readability that the readability formulas purport to measure.

Readability formulas are gathering in popularity, and in places that matter:

> "Today, reading experts use the formulas as standards for readability. They are widely used in education, publishing, business, health care, the military, and industry. Courts [in the USA] accept their use in testimony." (DuBay 2007, p. 5)

Organisations in the USA have been successfully sued by plaintiffs claiming that they have been disadvantaged by an inability to understand certain public documents.[3] To protect themselves against such litigation, many bodies—commercial and government—now have specific guidelines on the minimum readability required of their public documents. The result is that what was once an academic curiosity in cognitive science has been turned into a tool for profit-making. Companies abound that, for a fee, will help you improve your readability score. And to know if your readability has improved you need a test. So, just like intelligence tests, readability tests have become mainstream. The question of whether a better readability score translates into better readability is no longer considered a mere hypothesis. If Microsoft Word gives you a readability score, then readability formulas must, surely, have the imprimatur of scientific rigour. Or so many think.

Just as there is a variety of intelligence tests, there is a variety of readability tests. In fact, the history of readability

---

2. From E. Dale and J. S. Chall, "The Concept of Readability", quoted in William H. DuBay, *Smart Language: Readers, Readability, and the Grading of Text*, Impact Information, Costa Mesa, CA, 2007, p. 6.

3. For example, Tampa General Hospital and the University of South Florida paid a US$3.8 million settlement to a group of people who claimed that a consent form they signed exceeded their reading ability. Cited in DuBay 2007, p. 2.

research is littered with over 200 readability formulas, all of which appear to be *text-based*: that is, readability is considered a function of the text being read and has nothing to do with the reader. These tests consider as important such features as the average number of words per sentence, the average number of syllables per word, the number of single-syllable words, the number of polysyllables, the number of words not on some predetermined list of so-called *easy* words, and the like. For example, the Gunning Fog Index measures, in a sample of 100 words, the average number of words per sentence and the number of words of more than two syllables, and the Simple Measure Of Gobbledegook (SMOG) measures the number of words of more than two syllables in a sample of 30 words.

## The Flesch reading ease score

Probably the most influential of all the readability formulas is the Flesch reading ease formula, named after Rudolf Flesch. This is the formula Flesch came up with:

$$\text{Reading ease } (RE) = 206.835 - 84.6s - 1.015w$$

where $s$ is the average number of syllables per word and $w$ is the average number of words per sentence (Flesch 1948). In most cases, the value of $RE$ will fall between 0 and 100: the higher the value, the more readable the text (or so the theory goes).[4]

The Flesch reading ease formula (hereafter FREF) is behind the main readability statistic in Microsoft Word. It has also been tweaked for special uses (such as in the US Navy Readability Indexes). It also provides the raw input for another of the readability statistics that Microsoft Word generates: the Flesch–Kincaid Grade Level (which simply maps ranges of readability scores to particular levels of schooling in the US education system). Because of its special influence, most of our comments below will be directed at the FREF (although it

---

4.   "...on a scale between 0 (practically unreadable) and 100 (easy for any literate person)". Flesch 1948, p. 229.

should be clear that most of what we say will apply equally to any formula that derives a measure of readability from the properties of text alone).

At first glance, the FREF appears to have a degree of plausibility. It is undeniable that very long sentences *are* difficult to digest. By the time you have reached the end of a multi-clause sentence of, say, 50 or more words, you are often struggling to remember what you read at the start of the sentence. So it is difficult not to argue that the longer the sentence, the less readable it is.

But there is a fallacy lurking here that we need to be wary of. The undesirability of something doesn't imply the necessity of its opposite. Just because boiling hot food is not desirable, it doesn't follow that food should be served icy cold. Likewise, from the fact that long sentences are difficult to read, it doesn't follow that maximum readability demands the *shortest* possible sentence. A string of two- or three-word sentences is hardly likely to be maximally readable.

> The cat shook. It sat. It licked. It hissed. Then it slept.

is more difficult to absorb than the following longer version:

> The cat shook and sat. It licked and hissed and then it slept.

As a well-regarded style manual notes: "a string of short sentences can be irritatingly abrupt" (*Style Manual* 2002, p. 41). But the FREF gives a higher score the shorter the sentences. In other words, it encourages us to write very short sentences.

Now consider the importance of syllable count. On the face of it, polysyllabic words are more challenging than monosyllabic words. Most of us scratch our head, or disrupt our reading while we reach for a dictionary, when trying to read a medical book or book on philosophy. So a measure that reduces readability as syllable count increases, as Flesch's formula does, has *prima facie* plausibility.

But the relationship between readability and syllable count is superficial. It fails to acknowledge the influence that frequency of use has on a word's readability. There are numerous two-syllable words that are far more understandable than one-syllable words simply because we use them more often and have used them since childhood. For example:

> Mummy always washes the dishes after breakfast every morning

is obviously more readable than the equally long sentence:

> Electrons jump a level when hit by a photon

even though the former has a higher syllable count (18 as opposed to 13).

Indeed, a sentence can be short, with only monosyllabic words and yet be entirely obscure to the reader. For example:

> The work done was five ergs.

To many this sentence will be meaningless. There is no comprehension, no understanding. Only the scientifically-minded is likely to know that *erg* is a measure of *work* (in much the same way as *litre* is a measure of *volume*). But *The work done was five ergs* has the same number of words and syllables as the eminently readable *The cat sat on the mat*. By the FREF they are equally readable. Conversely, consider a tongue-twister like *hyperglycaemia*. This six-syllable word might baffle a healthy person, but to someone with diabetes, its meaning is crystal clear. These examples suggest that readability is intimately tied to conceptual familiarity (which is hardly a breathtaking discovery). If so, then text-only attributes—such as word length and syllable count—are not as important as Flesch believed.

On closer inspection the Flesch formula is seen to miss much of what makes a piece of writing readable. Unconventional grammar obviously gets in the way of understanding. It hinders readability. But it can still gain a high Flesch readability score. For example:

> Sat the mat the cat on

is a grammatically flawed sentence. Encountering it would frustrate our reading. And yet it has the same number of words and syllables as *The cat sat on the mat*.

In fact, the FREF necessarily gives the same readability score however you re-arrange the words in a sentence: with grammar and sense in mind or otherwise. For example,

   The on mat the sat cat

has the same number of words and syllables as *The cat sat on the mat* and thus should be equally readable on Flesch's view of readability. Indeed, it gets the maximum score: 100.

Punctuation, too, obviously affects readability. For example, if you write *Have a good holiday* when you should have written *Have a good holiday?*, then you are very likely going to confuse or mislead the reader. But the FREF gives these sentences the same score.

The FREF also gives the following sentences the same score:

   Their numbers can be estimated using an airborne particle detector.

   Their numbers can be estimated using an airborne-particle detector.

Omitting the hyphen in compound adjectives can create ambiguity if the context cannot clarify the intended meaning. (And the FREF, of course, is blind to context.) Is the detector in question a detector of cosmic rays tethered to a meteorological balloon floating in the stratosphere? Or is it a detector of carbon particulates fixed to the roof of a building?

The way text is styled can also affect readability, and yet style cues are ignored when all that is being considered is sentence length and syllable count. Consider this sentence:

   You must see the film My time before it ends.

This sentence is ambiguous because the title of the film is not clear. Is it *My time* or *My time before it ends*? But the FREF makes no judgement here. Too bad if I interpret it the wrong way and later encounter cognitive dissonance as a consequence.

Now consider non sequiturs: sentences that begin down one path and end down another. For example:

   Unlike many other areas of business where errors can be adjusted at a later date, employees immediately notice errors in their pay cheques

Whatever readability score this sentence gets, its meaning is impossible to determine. We can at best guess it. Indeed, it is easy to concoct non sequiturs that score the highest possible readability score, such as:

> When the cat sat on the mat, the square root of nine was three.

Any formula that attributes maximum readability to that sentence is undoubtedly faulty.

And what of vagueness? What meaning can we effortlessly attribute to *If there had been no activity for a long time, we applied heat to the solution* and *The current "seeped" through the substrate*? Imprecision, vagueness and the lazy use of near-enough quotes are obstacles to readability and yet the FREF is wholly oblivious to them.

Transition words are words having two equal or near-equal primary meanings. They are words that are part-way through a transition from one primary meaning to another. One meaning is waxing while another is waning. For example, a *villain* was once a yokel but is now a scoundrel. A *girl* was once a young child (of either sex) but now is a female child. The words *villain* and *girl* underwent a transition, and while they were in transition, their use could have been ambiguous. And thanks to the internet, and the democratisation of publishing it enables, we now have more concurrent transition words than ever before. An example is *disinterested*. Its strongest meaning not so long ago was *impartial* and *unbiased*; nowadays we see it used just as often to mean *bored* and *uninterested*. And does *regularly* still primarily denote *periodically*? Or is its use leaning more towards *frequently*? What is a transition word at one time might not be a transition word at another time. But when a word is in transition, its use is likely to be ambiguous if the context is of no help. Obviously, a readability formula that concentrates on textual statistics and ignores semantics will fail to detect the ambiguity posed by a transition word.

How many syllables comprise a word is not always clear cut. The number depends on how the word is pronounced, not how it is written, and this, of course, can vary between the different English languages. For US speakers, *temporary* has four syllables and *medieval* three; but for many others the syllable count is three and four respectively. Other words with varying syllable counts are *extraordinary, comparable, gaseous, medicine, laboratory*, and there are many more. Any formula that emphasises the importance of syllable count but considers only written language rather than spoken must make assumptions about how words are pronounced. Such assumptions obviously introduce further imprecision into the calculation.

To sum up: there is something fishy with Flesch.

## Correlation, volatility and validation

The arguments in the previous section should cast doubt on any claim that sentence length and syllable count—the only variables considered by the Flesch reading ease formula (FREF)—*define* or *cause* readability. Readability—the ease with which writing can be understood—is far too complex a notion. But could the score given by the FREF still be a good *indicator* or *predictor* of readability? An analogy: we do not define the concept of *temperature* in terms of the height of mercury in a thermometer. Rather, temperature is defined as the degree of hotness or coldness (and sometimes in terms of the average kinetic energy of the particles in a body). Nonetheless, the height of mercury in a thermometer has been found to be a very reliable indicator of temperature (at least on human scales). Indeed, the correlation between temperature and mercury level is as strong as can be. So, might textual statistics—sentence length and syllable count—be a good indicator of readability even though readability cannot be defined in terms of them? If so, a writer might reasonably use the FREF to evaluate the relative readability of various drafts of a document even though they might not write with textual statistics as their overriding guide.

This is indeed the claim of contemporary proponents of the textual analysis of readability. There is now widespread admission that many factors—not just sentence length and syllable count—contribute to readability and that writers should acknowledge those features when they are writing.

But despite this admission, many proponents argue that textual statistics are still the *best* indicator of readability, that if you calculate the correlation between other features of readability and comprehensibility the value you get is no better than if you calculated the correlation between textual variables and comprehensibility. In other words, an analysis of sentence length and syllable count is just as good as, but far simpler than, more complicated analyses.

> "Critics of the formulas … rightly claim that the formulas use only 'surface features' of text and ignore other features like content and organization. The research shows, however, that these surface features— the readability variables—with all their limitations have remained the best predictors of text difficulty as measured by comprehension tests." (DuBay 2007, p. 79)

But *best* does not imply *good*. At one time, the *best* way we had of estimating the number of stars in the universe was to look at the night sky and count them. But that, obviously, was not a very *good* technique. So how might we determine that the FREF gives a *good* way of indicating or predicting readability?

Any credible psychometric test must have two important attributes: *reliability* and *validity*. A test is reliable if it gives the same or similar result each time it is applied to the same subject. Obviously, the FREF is a very reliable test. Every time it is applied to the very same piece of text, it will give the very same score. What of validity? A test is considered valid if it actually measures what it purports to measure. This is usually determined by comparing the results the test gives against those given by an independent test widely accepted as being a good measure of whatever is being tested. A strong correlation between the results is considered to give the test in question validity.

So is the observed correlation between readability *as independently measured by reputable tests* and readability *as measured by the FREF* strong enough to warrant its use as a valid indicator or predictor of readability? In a typical readability research project, participants are given a number of texts and their comprehension of these texts is assessed. Two methods are widely used:

- Participants are given a number of questions about each text, and the number of correct answers supplied is used to assign a level of difficulty to a text.
- Participants undergo a cloze test whereby they fill in words that have been deliberately omitted from the texts. (In a typical cloze test, every fifth word is omitted.) The number of correct words added to a text is used to assign a level of difficulty to that text.

Once each text in a bank of texts has been graded, the relevant textual statistics in each are calculated and fed into a readability formula. Researchers then determine the correlation between the level of difficulty of a text as determined by a comprehension test (or cloze test) and the score given by the readability formula.

Correlation is simply a measure of how one variable changes when another variable changes. The most widely used formula for determining what is called the correlation coefficient, $r$, produces a value between $-1$ and $+1$. If $r = +1$, when the value of one variable is high, the value of the other variable is also high; and when it is low, the other variable is correspondingly low. If $r = -1$, when the value of one variable is high, the value of the other variable is low, and vice versa. Values between $-1$ and $+1$ indicate a less than perfect correlation, that is, the relationship between the two variables is loose and it is impossible always to infer with confidence what will happen to one variable when the other variable changes. And if $r = 0$, the two variables are completely independent: when the value of one is high, the value of the other is sometimes high and sometimes low, and equally so.

So how do the readability values generated by the FREF correlate with the readability levels determined by comprehension and cloze tests? In his paper introducing the FREF, Flesch reported a correlation coefficient of 0.7047 (Flesch

1948, p. 225). Follow-up studies have found widely varying correlations. Some have found a result similar to Flesch, and others reported lower results: 0.64 and 0.5 (Selzer 1983, p. 75). In a 1998 study of tourism texts, Woods found a correlation between text difficulty as assessed by cloze tests and text difficulty as determined by the FREF of just 0.13 (Woods et al. 1998). This is close to indicting that there is *no* predictive validity at all. The authors likewise found no high correlation between cloze scores and scores on a number of other popular text-based readability formulas (FRY, SMOG and FORCAST). They also noted that the four readability formulas gave widely inconsistent results, with some formulas scoring a text as considerably difficult when others scored it as relatively easy. They concluded that:

> "The readability tests examined in the present study gave very inconsistent results and none of the tests did a very good job at predicting readers' responses [to the cloze tests] … The results do not support the use of the readability tests analysed in this study (FRY, SMOG, FORCAST or Flesch's Ease of Reading test)."
> (Woods et al. 1998, p. 58)

Let's stay with the Flesch correlation for the moment and ask whether 0.7047 is strong enough to validate the FREF? A number of theorists have argued that it is not (Chall 1958). It might be true that such a value tells us that an inference from a high score on the FREF to high comprehensibility has a greater probability of being correct than incorrect; but it is also true that this inference will sometimes be incorrect. Still, doesn't a correlation coefficient of 0.7047 give us a *reasonable* degree of confidence that the Flesch reading ease score is useful?

It will be instructive to draw a distinction between *absolute readability* and *relative readability*. To ask whether a low Flesch score is a good indicator of low readability and a high Flesch score a good indicator of high readability is to ask whether the FREF is a good indicator or predictor of *absolute* readability. To ask whether the readability of a document is improved from one draft to the next if the Flesch score of the latter draft is higher than that of the former is to ask a question about *relative readability*. These concepts are quite different. If the FREF is a good indicator of absolute readability, it will have a high *predictive validity*. If the FREF is a good predictor of readability variation between drafts, it will have a high *comparative validity*. But, as we're about to see, low comparative validity is compatible with what is deemed by Flesch and his supporters to be high predictive validity. This might seem paradoxical, but only until we question whether a correlation coefficient of 0.7047 is really enough to give the FREF high predictive validity.

Let's start by considering whether the correlation coefficient Flesch found implies significant comparative validity. In other words, does a correlation coefficient of 0.7047 give us confidence to conclude that draft 2 of our document is more readable than draft 1 if the Flesch score of draft 2 is higher than that of draft 1? Table 6.1 below suggests that it does not.

Table 6.1: Correlation and volatility

| Validation | | | 1st draft | | 2nd draft | | F change | M change |
|---|---|---|---|---|---|---|---|---|
| Flesch | Master | | F1 | M1 | F2 | M2 | | |
| 25 | 35 | | 25 | 35 | 30 | 30 | Up | Down |
| 35 | 20 | | 35 | 20 | 25 | 35 | Down | Up |
| 42 | 25 | | 42 | 25 | 35 | 42 | Down | Up |
| 42 | 40 | | 42 | 40 | 40 | 62 | Down | Up |
| 65 | 50 | | 65 | 50 | 50 | 52 | Down | Up |
| 55 | 45 | | 55 | 45 | 45 | 50 | Down | Up |
| 60 | 65 | | 60 | 65 | 71 | 51 | Up | Down |
| 65 | 50 | | 65 | 50 | 50 | 60 | Down | Up |
| 68 | 50 | | 68 | 50 | 56 | 70 | Down | Up |
| 75 | 80 | | 75 | 80 | 80 | 70 | Up | Down |
| 30 | 30 | | | | | | | |
| 25 | 35 | | | | | | | |
| 35 | 42 | | | | Predictive validity = | 0.705083 | | |
| 40 | 62 | | | | | | | |
| 50 | 52 | | | | | | | |
| 45 | 50 | | | | Volatility = | 100% | | |
| 71 | 51 | | | | | | | |
| 50 | 60 | | | | Comparative validity = | 0 | | |
| 56 | 70 | | | | | | | |
| 80 | 70 | | | | | | | |
| Correlation = | 0.705083 | | | | | | | |

Imagine that the first draft of each of 10 documents is subjected to the FREF and they get the reading ease scores listed in the **F1** column in table 6.1. Suppose further that the same documents are subjected to the FREF at the end of their next draft. The FREF scores for that draft are listed in column **F2**. Suppose now that the 20 drafts—the 10 first drafts and 10 second drafts—are subjected to a master test of readability (a cloze test, comprehension test or whatever type of test Flesch used to validate his formula). The score for each draft is listed in the **Master** column in table 6.1 beside its corresponding Flesch score. Importantly, we have deliberately chosen scores so that the correlation between the Flesch score and the master score turns out to be almost identical to the correlation Flesch considered sufficient to support his reading ease formula, namely 0.7047. (The scores in table 6.1 yield the slightly better 0.705083.)

Now to a surprising feature of these scores. For every Flesch score that went up between drafts, the actual readability (as measured by the master test) went down, and for every Flesch score that went down the actual readability went up. To see this, compare each **F1** score against its **F2** score and note the difference between the corresponding **Master** scores (**M1** and **M2**). Every change in a Flesch score brought about a change in readability *in exactly the opposite direction* to what the FREF would predict. (You can see this easily in the **F change** and **M change** columns: for every *Up* there is a corresponding *Down*, and vice versa).

Let us introduce the term *volatility* as the measure of the degree of directional variability in a sample of paired scores. Volatility occurs when one value in a pair increases (or decreases) and the corresponding value in a related pair of scores decreases (or increases). Maximum variability (measured on a scale of 0 to 100%) occurs when *no* increase (or decrease) in any pair in the sample is matched with an increase (or decrease) in a related pair. That's the situation illustrated in table 6.1. Thus the volatility of 100: maximum volatility.

Flesch's correlation of 0.7074 is certainly better than 0 so, on the face of it, it should give you some confidence that a low Flesch score does indicate low readability. If you are a fan of the FREF and got a low score for a draft of a document you are writing, you would no doubt consider polishing it up and checking it again against the FREF in the hope of getting a higher readability score. But this is where you can become unstuck. For as table 6.1 shows, Flesch's predictive validity is compatible with *maximum* volatility. If you get a higher Flesch score for your second draft, there is absolutely no reason for you to be confident that the actual readability has increased.

The confidence that a writer should be seeking is what I've called *comparative validity*: the likelihood that a higher Flesch score indicates a higher readability (and vice versa). When comparative validity is 1, volatility is zero and all increases in Flesch scores correspond to increases in actual readability (and vice versa). Writers can then be confident of improving the readability of their drafts if the FREF gives a higher score. But, as table 6.1 shows, the comparative validity is zero. No movement in Flesch score corresponds to an equi-directional movement in readability score. Writer confidence? Zero. A predictive validity of 0.7047 has misled us about the extent to which the FREF is useful.

You might retort that we concocted the scores in table 6.1. We did, but that is not relevant. All we have from Flesch is a correlation coefficient of 0.7047. This is meant to be an indication of the predictive validity of the FREF. The point is that the predictive power of a formula with a seemingly useful correlation coefficient of 0.7047 is compatible with indeterminate intra-score predictability (that is, maximum volatility). It is simply not enough to know that the correlation between two sets of measurements is 0.7047 to be confident that a rise (or fall) in a value of one measurement will more often than not correspond to a rise (or fall) in the value of the correlated measurement. The correlation is compatible with a *comparative* validity of zero. But this is the very sort of validity writers need to know and be confident of if they are going to rely on the FREF to judge the relative readability of drafts.

Flesch scores are used in the USA to determine the suitability of books for certain school grades. Our argument so far has had to do with comparing the Flesch scores on various drafts of the same document. But within the confines of our argument, there is no relevant difference between comparing draft 1 against draft 2 of the same document and comparing book 1 against book 2. If an educator finds that the Flesch score for book 1 is greater (or less) than the Flesch score for book 2, a predictive validity of 0.7047 should give them no confidence that the readability of book 2 is greater (or less) than book 1. At best, such predictive validity suggests that a low Flesch score is associated with a low readability level and vice versa. But even that supposition is questionable, as table 6.2 below suggests.

Table 6.2: Comparative v. predictive validity

| Validation | | | 1st draft | | 2nd draft | | F change | M change |
|---|---|---|---|---|---|---|---|---|
| Flesch | Master | | F1 | M1 | F2 | M2 | | |
| 56 | 35 * | | 56 | 35 | 50 | 30 | Down | Down |
| 40 | 20 * | | 40 | 20 | 57 | 35 | Up | Up |
| 26 | 25 | | 26 | 25 | 42 | 42 | Up | Up |
| 43 | 40 | | 43 | 40 | 62 | 62 | Up | Up |
| 70 | 50 * | | 70 | 50 | 51 | 52 | Down | Up |
| 56 | 45 | | 56 | 45 | 49 | 50 | Down | Up |
| 45 | 65 * | | 45 | 65 | 52 | 51 | Up | Down |
| 55 | 50 | | 55 | 50 | 59 | 60 | Up | Up |
| 49 | 50 | | 49 | 50 | 68 | 70 | Up | Up |
| 79 | 80 | | 79 | 80 | 61 | 70 | Down | Down |
| 50 | 30 * | | | | | | | |
| 57 | 35 * | | | | | | | |
| 42 | 42 | | | | | | | |
| 62 | 62 | | | | | Predictive validity = | 0.70477 | |
| 51 | 52 | | | | | | | |
| 49 | 50 | | | | | Volatility = | 30% | |
| 52 | 51 | | | | | | | |
| 59 | 60 | | | | | Comparative validity = | 0.7 | |
| 68 | 70 | | | | | | | |
| 61 | 70 | | | | | | | |
| Correlation = | 0.70477 | | | | | | | |

Table 6.2 gives another dataset where the correlation is close to Flesch's 0.7047. In this case, the volatility is 30%, which gives us a probability of 0.7 that an increase (or decrease) in a Flesch score between drafts will correspond to an increase (or decrease) of readability. That's better than the previous dataset, you might think.

But table 6.2 also suggests that a correlation of 0.7047 or thereabouts is not even a very good *predictor* of readability. You can see this by noting that in 6 out of the 20 drafts, the difference between the Flesch score and the corresponding readability score is at least 20 points. (These are marked with an asterisk.) Given that the Flesch scale is only 100 points wide, a 20-point discrepancy is something to worry about. (In the USA, where Flesch scores are used to select the suitability of books for certain school grades, a discrepancy of 20 Flesch points could see a book suitable only for Year 4 students set on a Year 3 curriculum or vice versa.) Thus, with Flesch's observed correlation, we could be wrong 30% of the time in assigning a readability measure to a document based on the FREF. Indeed, it is possible for a predictive validity of 0.7047 to co-exist with 100% of scores being at variance by at least 10 points. It seems, then, that the predictive validity of the FREF is not really as strong as the formula's proponents have assumed.

Perhaps in recognition of the fact that a correlation coefficient of 0.7047 is compatible with significantly volatile data, current proponents of readability formulas have begun tempering their enthusiasm for them, now admitting that they can only be rough guides:

> "[Readability formulas] ... are not perfect predictors. They provide probability statements or, rather, rough estimates ... of text difficulty." (DuBay 2007, p. 110).

But whatever correlation coefficient is obtained, the typical validation methodology that readability researchers adopt *overstates* the actual coefficient. The reason is that the methodology excludes certain types of material from being tested. And if these types weren't excluded, the testing would be of little value (as we'll now show).

Any study of scientific merit does not arbitrarily limit its data sampling. If you want to establish that *all* bodies fall with the same rate of acceleration, you do not limit your experiments to, say, metal objects. But the problem with relying only on the results of comprehension or cloze tests to determine the correlation between reading difficulty and readability statistics is that the data sample is necessarily limited. This is because those who devise a comprehension or cloze test can only use texts that are fully comprehensible (at least to them), otherwise they would not be able to determine if an answer provided by a testee is the correct answer. Indeed, they would find it difficult, if not impossible, to devise sensible questions in the first place.

But to avoid skewing the results, the data sampled must be extended to include texts that are *partially* comprehensible and even *in*comprehensible. Assessing the *degree* of comprehension a text offers would be an extraordinarily difficult challenge. Fortunately for our argument we can side-step that issue and concentrate, instead, on incomprehensible texts. These are much easier to find. A politician's speech might be a good place to start. But incomprehensible writing can also be concocted. We could write passages that are hopelessly ambiguous, or hopelessly vague. It is much easier, though, just to concoct nonsense strings, such as *The door is in love* and *Honesty is the largest integer less bilious than the smartest flooring wart.*

Samples of nonsense strings don't need to be tested for comprehension before they are subjected to the FREF. We know, by definition, that they have a comprehension value of zero (for they are nonsense). Now if the FREF is a valid indicator of readability (and thus of comprehension) then nonsense strings should get an FREF score of zero. (Recall that FREF scores range from 0 to 100, with zero indicating that the text is incomprehensible and 100 that it is fully comprehensible to any literate person.)

However, it should be clear that there can be no correlation whatsoever between the comprehension scores of nonsense strings and corresponding FREF scores. This is because we can concoct any number of nonsense strings with few words and few syllables (which would score high on the FREF: the first example above—*The door is in love*—scores 100) and any number of nonsense strings with many words and many syllables (which would score low on the FREF: the second example above scores 46.6). And if we are conducting the experiment scientifically, we need to include all types of nonsense strings: short, long, monosyllabic and polysyllabic. Thus a zero actual comprehension score can be matched to any value in the 0–100 range of FREF scores. *This clearly indicates a correlation coefficient of zero.*

So, for maximally comprehensible texts, the best correlation coefficient that testers can find between comprehension scores and FREF scores is 0.7047. And the only correlation coefficient possible for incomprehensible texts is zero. It follows that if correlation testing were to include samples of all types of texts—comprehensible and incomprehensible— the real correlation coefficient must be even less than the already unconvincing 0.7074 that Flesch reported (and perhaps even less than the insignificant 0.13 that Woods and her co-researchers found: see page 87).

## The irrelevance of readability formulas

Let's return to the fact that studies to establish a correlation between scores on comprehension tests and scores on the FREF must use texts that are maximally comprehensible to the testers. If they didn't, the answers testees give would have no value in determining a text's degree of difficulty. But how will those devising a test know that a candidate text is fully comprehensible? They can't subject it to a comprehension or cloze test to assess whether it is suitable to be subject to a comprehension or cloze test. Nor can they subject it to a readability formula, since the validity of the formula is the very thing they are trying to prove. So for testers to determine whether a text is fully comprehensible there must be some other criterion available for them to use, that is, a criterion other than a score on a comprehension, cloze or Flesch test.

So the methodology boils down to this: you start with a reputable criterion or test of comprehensibility (which, on our earlier definitions, would be a test of readability) to select fully comprehensible texts for a comprehension or cloze test. You then use the results from the comprehension or cloze test to validate a readability formula, finding that the best correlation coefficient you can find is about 0.7 (which is over-stated due to sampling limitations). That is, you start with a criterion that must give a good measure of readability and use it to validate a formula that can at best give just a rough estimate of readability. *Wouldn't it be so much better just to use the initial criterion as our test of readability and forget about readability formulas altogether?*

Paradoxically, some proponents of readability formulas come close to adopting this view. For example, after a long and favourable consideration of many formulas, the most that DuBay can say is:

> "When adjusting a text to the reading level of an audience, using a formula gets you started, but there is still a way to go. You have to bring all the methods of good writing to bear."[5]

But if we still have to bring all the methods of good writing to bear—that is, to concentrate on all those features of text that are essential for maximal readability—why don't we just concentrate on those features and forget about the additional task of applying the FREF?

DuBay's quite sensible advice would not be welcomed by those who argue that for workaday uses we need a simple proxy measure of readability, such as a text-based formula, because directly assessing readability is just too difficult and time-consuming. I suspect that those who take this line may have gone through schooling during that Dark Age of English

---

5.  DuBay 2007, p. 112. This seems to contradict DuBay's earlier claim that when you consider all the features of a text, textual statistics are still the *best* predictor of readability.

language teaching: the latter quarter of the twentieth century. Assessing readability might be a multi-faceted task, but it is not especially difficult. Any reputable language manual is a good start. Indeed, many of the examples given in the previous section to discredit the conceptual linking of readability and textual statistics can guide us here, enabling us to distil some of the general features of readability. And a moment's reflection should make it clear that familiarity, clarity, neutrality, conceptual lightness and consistency affect readability (qualities ignored by purely text-based formulas). Devising a text-based, reader-independent algorithm that will assign an objective value to each of these qualities is highly unlikely, and probably impossible. (For a start, familiarity is as much a function of the reader as of the text.) It may be time, then, to dispose of text-based readability formulas on the scrap heap of over-zealous quantification. There is no shortcut to determining the quality of your writing. You need to sift out the ambiguity, vagueness, foreignness, verbiage and the like on your own (or with the help of an editor). There is simply no app with any claim to scientific merit that will do it for you.

# 7: Is structured authoring a paradigm shift?

What do technical writers need to know in order to gain sustained employment in the technical writing profession? Contrary to the pronouncements of some technology evangelists, knowledge *of* XML (as distinct from knowledge *about* XML) is not necessary. It's true that much software these days is glued together with XML, and true too that some technical writers produce XML-tagged content. But most XML-tagged content is destined for the information technology (IT) domain. And while this domain is certainly an important source of revenue for technical writers, what some technology evangelists overlook is the fact that many technical writers work in non-IT domains: medical equipment, heavy machinery, mining and so on.

But isn't an authoring methodology separate from any specific domain, being a general approach that could be applied regardless of domain? In other words, isn't XML-based authoring suitable whether you are working in IT or in heavy machinery? Well, yes; but recall that the question is whether knowledge *of* XML (as distinct from knowledge *about* XML) is necessary. One can engage in XML-based authoring without needing to know the slightest thing about XML prologs, namespaces, IDREFs and the like. A parallel: do you really need to know the ins and outs of C++ in order to author a document in Microsoft Word or Adobe FrameMaker? Those who created the authoring infrastructure did, but not the day-

First published in *Words*, vol. 1, iss. 4, November 2009

to-day authors. It's much the same in technical writing. There are documentation technicians who set up the authoring infrastructure for XML-based authoring—create schemas, DITA specialisations and the like—and there are those who use that infrastructure to author. The former need to know the nuts and bolts of XML; the latter do not.

So those coming new to technical writing need not worry too much if they know little of the nuts and bolts of XML. But what about the content-driven authoring *technique* that produces the XML (or XML-tagged content)? Isn't the move to content-driven authoring from format-driven authoring something new, implying a whole new approach to authoring? Isn't it a *paradigm shift*, something that writers old and new need to embrace?

Some technology evangelists certainly think so:

> "The evolution of content creation from format-driven publishing to structured authoring is a paradigm-shifting … transition." (O'Keefe S. 2008, p. 27)

First, what is a *paradigm shift*? The concept was introduced by Thomas Kuhn in *The Structure of Scientific Revolutions* (Kuhn 1962). Kuhn applied the concept to science. It comes about when a new discovery or new theory, in explaining something that had been considered anomalous by current theories, forces a radical or revolutionary change in scientific worldview. The shift from a geocentric to a heliocentric view of the universe is a paradigm shift. So too is the shift from creationism to evolution, and from classical mechanics to quantum mechanics.

The term has taken on a life of its own and is no longer applied only to scientific revolutions. Thus a move from a Keynsian view to a monetarist view is considered a paradigm shift in macroeconomics. Indeed, the term today seems to mean nothing more than a radical change in thinking about, or doing, something, regardless of domain.

So, is the move to structured authoring—the move to content-driven authoring as opposed to format-driven authoring—a paradigm shift? Does it constitute a radical change in the way technical writers write?

Firstly, does the question really matter? Isn't this merely quibbling over semantics? In one sense, yes; but it's more than just semantics when a term misleads, especially when it misleads those new to our profession or those wanting to break into it. To call a methodology *paradigm shifting* is to imply that the methodology it has supplanted, or is supplanting, is outmoded, erroneous, ineffective and the like: that is, it implies that the new methodology constitutes some intellectual advance on the old. More importantly, it implies that our profession has seen the light and moved across to structured authoring. Thus, for someone wanting to break into technical writing, they would need to master structured authoring.

But this is just not true. Very few technical writing projects adopt a structured authoring approach. For most projects, such an approach would be like taking a sledgehammer to a walnut. Large organisations producing multi-lingual deliverables along multiple delivery channels can gain advantages from a structured authoring approach. But for the common-or-garden projects most of us work on, there would be no gains from a structured authoring approach. So, despite the hoopla of some technology evangelists, those new to our profession, and those wanting to break into it, needn't feel that they must master structured authoring if they are to have any chance of advancing as a technical writer.

But give the new methodology time, some might say. This is all new, and it will take time for it to become as commonplace as today's predominantly format-driven authoring. But structured authoring is not a new methodology at all. It has been around since the 1980s, when Standard Generalised Markup Language (SGML) first appeared, more than a decade before the birth of its child, XML. (XML is just a sub-set of SGML.) The hype and hoopla that greeted the arrival of SGML caused many technical writers to attempt to master SGML, and many bought *FrameMaker+SGML*, one of the first structured authoring tools on the market. But SGML has now largely faded away, and it is arguable whether modern approaches to structured authoring will overcome all of the difficulties — including wide-scale irrelevancy — that befell SGML. (SGML can still be found in some organisations, but very few technical writing projects worldwide depend on SGML.)

But let's return to the issue of whether content-driven authoring constitutes a paradigm shift, a radical new way of authoring. The old paradigm is *format-driven* authoring, where the predominant concern, apparently, is the *appearance* of our text rather than the types of building blocks that make up the content. The new paradigm has us concentrating solely on content and ignoring formatting (or at least relegating it to the secondary task of applying some stylesheet or coded transform *post-drafting*).

But is this really a new approach to authoring? Does any writer, can any writer, really adopt a format-driven authoring methodology to the exclusion of a content-driven authoring methodology? Suppose, for example, that I am writing a scientific paper describing the results of some research. Do I really say to myself: first, I shall start with some heading 1 text, move on to a heading 2 text, choose a smaller-than-normal body text format (maybe with some left and right indents) for the next paragraph, then add some more heading 2 text, and then some standard body text, and so on and so on. Of course not. We *naturally* think in terms of content: I start with a title, then I write the authors section. Next I add an abstract (introduced with its own heading), followed by the introduction, then the materials and methods section, followed by the results section, the discussion section, the acknowledgements, appendices and finally the list of references. The whole paper is written in content chunks, not format chunks. We will certainly format the content chunks (and the sub-chunks: headings, lists and so on). But we primarily think of the paper we are writing as composed of chunks of content or topics. Formatting is always secondary. And this is not new.

It is exactly the same with technical writing (and with any form of declarative writing). Formatting is always an afterthought. The fore-thought is the content types that will be the building blocks of my document. I don't say to myself as I am about to begin the steps in a procedure that I am choosing a list format. No, I say to myself that I am about to begin a procedure. This is a content type, not a format type. I may format it in a particular way—as a numbered list—but that is not the primary consideration. The primary consideration is that I am about to set out the steps to describe how a specific goal can be achieved. When I write a warning or caution, I am not saying in my mind that I am about to enter some bold text with an accompanying danger symbol. No. I say I am about to enter a warning or a caution. This is a decision about content. When I am writing a trouble-shooting section, I don't say that I am choosing a particular format type. No, I am choosing a particular content type. And so on. (This is the logic behind *boilerplate templates*, the most common type of template: here are all the sections, that is, major content blocks; now plug in your text.)

So, in a fundamental sense we have *always* been engaged in structured authoring, despite the relatively recent appearance of structured authoring tools. Far from being a paradigm shift in how we author, structured authoring methodologies are actually doing little more than mirroring, at last, the way we have always authored. *It is the tools that have changed; it is not how we author that needs to change.* The structural components (or elements) that we see in DocBook, DITA and the like are just a reflection of the way we *naturally* chunk our writing, how we build a document from the blocks that are its necessary constituents. If it's always been that way—and it has—then authoring that way cannot be a paradigm shift.

Strictly speaking, how we author in a structured authoring environment is a little different, but the difference is not such that a would-be or novice technical writer needs to be especially concerned. With the structured authoring that we naturally adopt but with a non-structured authoring tool at our disposal (such as Microsoft Word), we are free to construct the structure as we please and free, too, to apply any format to any paragraph (and any character) that we type. With the structured authoring that we naturally adopt but with a structured authoring tool at our disposal (such as Structured FrameMaker), we are free to apply whatever content type is appropriate wherever we have our cursor. This is determined by the content rules in the associated schema, document type definition (DTD) or element definition document (EDD). Formatting is another step: it is either applied via an associated style sheet or set out in the format rules specified in the associated EDD. But the principal difference is that with an unstructured authoring tool, you apply styles (aka formats) directly to the text you enter, whereas with structured authoring tools you directly apply content types (aka elements and their qualifying attributes) to the text you enter. Formatting is another step. With an unstructured tool, you select a style to apply to a chunk of text; with a structured tool you select a content type to apply to a chunk of text.

And what is so smart about this is that the structured authoring methodology exactly mirrors the way we author, and the way we have always authored. *This is no paradigm shift or quantum leap. Rather, the tools are catching up with us. It is not us who has to catch up with the tools.*

So don't be alarmed if you are new to technical writing and are confused by all the hype and hoopla about the need to learn XML and the need to embrace a new model of authoring. The model that many are apparently moving to is not new. It is a model that simply reflects the way everybody writes. If you have written anything at all—with quill, crayon, chalk or Microsoft Word—you have more than likely engaged in structured authoring: content first; format second. What *is* new is that modern authoring tools—from the advent of SGML onwards—can *enforce* a particular structure on a document. Those new to technical writing can learn how that can be done, if they wish. But not knowing how it can be done is no obstacle to advancing in the profession—especially given that so few technical writing projects insist on structured authoring.

# 8: Returning language back into the spotlight

Like relativity in physics, usability in documentation is a concept that simply can't be ignored. It colours—or should colour—every decision we make in designing and writing documentation. But like relativity, pinning down a useful definition of usability is no easy matter.

The International Standards Organization describes usability as "the extent to which a product can be used by specified users to achieve specified goals with effectiveness, efficiency, and satisfaction".[1] This gives some conceptual traction, but it lacks the necessary concreteness to make it immediately applicable.

A more concrete definition, and one more widely discussed, is based on the work of Gretchen Hargis and her colleagues in defining quality documentation. (Hargis et al. 2004) This view of usability (and quality documentation) has it that the information in documentation must be:

- easy to find
- easy to understand and
- easy to apply.

The linking concept here is provided by the common definition of quality as "fitness for use". Obviously, if a

---

1. International Standards Organization, Human-centred design processes for interactive systems, ISO 13407:1999. The definition is repeated in numerous ISO standards directed at technical communicators, such as ISO/IEC 18019:2007 and ISO/IEC 26514:2008.

First published in *Intercom*, June 2009, pp. 4–6

document is not fit for use, it lacks usability; and if it is fit for use, it has usability — at least to some degree. And it is plausible to judge that degree on how easy the information in it is to find, understand, and apply.

For information to be easy to find, there must be sufficient signposts in places where readers are likely to look. The two most likely places are an index and a contents list, with an index arguably the more important of the two in a document of more than a score or so of pages. A contents list, though useful to the occasional browser, simply hasn't the degree of granularity needed to help the typical reader: the time-poor, deadline-harassed person needing to know in a hurry how to do the particular task at hand. For such a reader, a lengthy document without an index would be deficient in usability.

Other aides in helping users find information easily include running headers and footers, cross-references and hyperlinks, lists of related tasks, breadcrumbs, and a full-text search facility (especially one that enables wildcard searching and Boolean filtering). All these features, to varying degrees, help readers find the information they are after, and thus contribute to the overall usability of documentation.

Once a user has found the information they are after, they need, of course, to be able to understand it. This is where language and usability intersect, and the influence of the former on the latter is the main topic of this paper. I will come back to it shortly.

The other pillar of usability is that the information, once found and understood, must be easy to apply. To that end, it must deliver what it promises. A procedure promising to explain how to set a timer on a video recorder might be easily found and well-written, but if it doesn't fully explain how to set a timer, and under all likely conditions, then it is less than maximally usable. Moreover, it should not cause the reader to back out of the procedure by introducing prerequisites in the steps rather than in the preamble, nor cause the reader to consult other sections of the user guide in order to complete the procedure they are working their way through. Thus the information must be relevant, accurate, comprehensive, and self-contained.

But let's return to the second pillar of usability: the information presented in documentation must be easy to understand. This is arguably the most important facet of usability in the documentation field. There may be a plethora of signposts directing a reader to the procedures they might need (and thus the information is easy to find) and each procedure may well cover all conditions and be self-contained (and thus score not too badly on the easy-to-apply scale), but if a reader has to struggle to understand the information presented to them, then the usability of the document is undeniably deficient.

But what is meant by easy to understand?

## Understandability and readability

One often hears the KISS principle extolled in technical writing circles: Keep It Simple, Stupid. Alas, the KISS principle is hoist with its own petard. It is just too simple to be of any use. Still, much effort has gone into providing simple measures of understandability, measures that, unlike the KISS principle, have some prima facie claim to scientific rigor. These are the so-called text-based readability formulas, the most well-known of which is the Flesch reading-ease formula (the maths behind the readability scores generated by Microsoft Word).

For a start, readability and understandability are often used interchangeably:

> "Readability means understandability. The more readable
> a document is, the more easily it can be understood ..."
> (Samson 1993, p. 58)

Hence readability formulas such as the Flesch reading-ease formula can be considered contenders for determining the usability of documentation (or at least that component related to ease-of-understanding).

But the Flesch reading-ease formula errs on the side of KISS-like simplicity. It takes as its input just two features of text:

average sentence length and average syllable count. Nothing about the reader is included, such as their familiarity with the concepts discussed. And many features of text that necessarily contribute to, or detract from, understandability are ignored: conventional grammar and punctuation, typographical cueing, contradiction, inconsistency, non sequiturs, ambiguity (especially that resulting from the use of transitional vocabulary), and many more. It is just far too easy to concoct a difficult, or even nonsensical, piece of text that scores well on the Flesch reading-ease formula. Short sentences and monosyllabic words do not understanding make.

To those who accept these limitations but argue that the Flesch reading-ease formula is still the best proxy measure of readability we have (DuBay 2007, p. 79), we can retort that best does not imply good. At one time, the best way we had of estimating the number of stars in the universe was to look at the night sky and count them. But that, obviously, was not a very good technique. Further, numerous studies have failed to reproduce the sort of validation correlation that excited Flesch—the correlation between Flesch scores and scores on independent comprehension tests—and any such correlation is necessarily inflated by ineradicable sampling bias. See "Can the quality of writing be measured?" starting on page 75.

We should not be fooled, then, into thinking that its use in Microsoft Word gives the Flesch reading-ease formula the imprimatur of scientific rigor. The formula is overly simplistic and offers little guidance in determining whether a piece of text meets any likely usability criterion.

## Understandability and communicative efficiency

We get closer to an understanding of *understanding* if we reflect on why we write, namely, to communicate. We communicate if we get our message across. But our success in getting our message across can be judged in degrees. We might achieve effortless communication: our readers get our message immediately, without any cognitive or emotional struggle. At the other end of the spectrum, we might fail completely: ambiguity, vagueness, conceptual denseness, and a host of other factors might block all attempts at deciphering our intended message. And in between are the readers who eventually work out what we mean, but only after some degree of struggle, or an encounter with more words than were necessary to get the message across.

Communicative efficiency captures the notion of ease-of-understanding far better than sentence length and syllable count. Efficiency entails effectiveness: obviously we need to get our message across if our communication is to be efficient. But it also entails that we get our message across with the least effort on the part of our readers. In other words, we should write with maximum economy, using language that is most familiar to our intended audience, and which has the least potential for distraction (which might arise, for example, if we engage the emotions of our readers with paternalistic or insensitive language, or if we use language inconsistently).

## Usability, words and the flight from technical writing

Ease of understanding, and thus usability, depends, then, on our writing exhibiting clarity, economy, familiarity, neutrality, and consistency. And thus it is impossible, in our field, to achieve maximum usability without a pre-eminent respect for language and for the words that are its building blocks. For we risk failing to get our message across if a careless choice of words leads to ambiguity, vagueness, bafflement, offence, or cognitive overload.

Words, then, should be at the centre of our professional concerns. And yet words and language can often seem of marginal concern to technical communicators. The threads on discussion forums, the articles published in our journals, and the marketing materials designed to attract students to our university courses, lean strongly toward tools, methodologies, and practices. Issues of language are often missing or downplayed.

Our obsession with broadening our profession's profile — apparent in the number of times we have changed our name — may have contributed to the drift away from appreciating the importance of words. We were once technical writers, and when we were, the importance of writing — of words and of language — was explicit. It needed no explaining. But we did, have always done, more than writing, and thus we felt a need to be called something else: technical communicators, content providers, end-user assistance professionals, information designers, and so on.[2]

But other professions are not so touchy about their name. Teachers do more than teach. They also act as playground monitors, sports-day referees, mentors, excursion leaders, and curriculum designers. But they still call themselves teachers. We do more than write, but, unlike teachers — and many other professionals — we have sought to change our profession's name to make what we do explicit.

In the process, we have ended up achieving the opposite: concocting names of such bland generality as to encompass many clearly distinct professions. (A journalist, graphic designer, and musician can all be seen as content providers; and a call-centre representative is also an end-user assistance professional.) We have failed to identify and differentiate ourselves by adopting names that drown out our particular, unique contribution. And in doing so we may have lost sight of the fact that writing is what most of us do most of the time (just as teaching is what most teachers do most of the time). We may be especially fond of tools and methodologies — and there is no harm in that; indeed some degree of tools expertise is essential — but expertise in XSL transforms, DITA, persona mapping, VBA macros, Framescript, wiki design and the like is of no use if our writing — our particular, unique contribution — fails to achieve its primary purpose: effortless

---

2.  This is considered in detail in chapter 1, "Technical writing: what's in a name?" starting on page 9.

communication. It is writing before all else, and that is so even if some in our profession spend all their working time doing things other than writing.

To its credit, our profession has always prized usability. We may not have always agreed on what it means, nor given due respect to the need to clarify its definition. But a moments reflection on why we do what we do, on the ISO definition of usability, and on the work of Hargis and her colleagues, should bring home the fundamental importance of language to our profession. Words are what make or break us. Our technical skills are secondary, and have always been secondary. Their relevance changes from year to year, version to version—unlike that of language. So if we are to continue our commendable respect for usability, we must return language to the spotlight. We must develop a passion for language that matches that of lexicographers. We must put down our prescriptive grammars and become scientists of linguistic flux. We must accept that being a users' advocate—which most of us do—requires immersion in the users' language. For what good is an attractive, well-structured document—even a well-crafted sentence, written once and re-used often—if it fails to deliver its meaning to the audience for which it was intended.

# 9:   The pseudoscience of Information Mapping

## Part 1: The 7 +/- 2 chunking limit

There is a view—especially prevalent in the technical writing profession—that information should only be presented to readers in small quanta or chunks. On the face of it, this is a sensible view. A sentence composed of three or more clauses usually breaches the limit of cognitive lodgment. It is just too long for most readers to absorb. Only by atomising a conglomerate of ideas into discrete sentences—that is, by chunking our ideas—do we minimise the effort our readers need to exert to understand what we have written.

We also chunk our writing into paragraphs. A paragraph enables a writer to present an argument, discussion, analysis or whatever in logical and accessible units:

> "There are two purposes for paragraphs, the one logical and the other practical. The logical purpose is to signal stages in a narrative or argument. The practical one is to relieve the forbidding gloom of a solid page of text."
> (Hudson 1993, p, 294)

Paragraphing is thus a form of chunking. Without it, readers would be presented with a form of *scriptio continuo*: undemarcated logic rather than undemarcated words. Such writing would hardly be inviting, nor likely to keep readers engaged. They would have to work out for themselves where each new topic began, causing them to drag anchor as they

First published, in three parts, in *Words*, vol. 3, issues 2–4, 2011 read. So chunking has been part of writing for a very long time. We chunk ideas into sentences, groups of related ideas into paragraphs, groups of related paragraphs into sections, and so on.

Traditionally, writers could choose the length of a sentence, a paragraph and a section. The amount of material they had about a particular topic, and the form of reasoning being adopted, might place limits on how long a paragraph or section might be. But these were limits imposed by logic and they varied from paragraph to paragraph. No-one—or at least no-one before the late 1960s—considered that there had to be an upper limit to the size of a paragraph or section no matter how much information the writer had to deliver. We might have deliberately avoided presenting readers with solid pages of text, but we did so without regard to any particular limit on the size of our chunks.

Today there are some who believe that we should limit our chunks to 7 ± 2 units of information, whether we are writing instructional materials, designing billboards or programming computer code. The source of this belief is usually attributed to an analytic literature review published by the American psychologist George Miller in 1956, the title of which is "The Magical Number Seven, Plus or Minus Two: Some Limits on our Capacity for Processing Information" (Miller 1956). Many writers were introduced to the notion of a chunking limit through training in what was once a popular writing method: *Information Mapping*. The method is still taught and thus the so-called magical span of 7 ± 2 still holds a spell over many writers. But is this science or flim-flam? Let's consider the Information Mapping method first before looking at Miller's often-quoted paper, the paper said to be the foundation on which the method was built.

## Information Mapping

The Information Mapping method was developed by the American political scientist Robert Horn. According to Horn:

"Writers should group information into small, manageable units ... A 'manageable unit' of information is one
consisting of no more than nine pieces of information ...
Rationale: Research suggests that people can best
process and remember no more than seven plus or
minus two pieces, or units, of information at one time ...
Therefore, a general guideline for a 'manageable unit of
information' is one consisting of 7 ± 2 pieces (also
referred to as the chunking limit) ... Writers should
create units of information that do not exceed the
chunking limit. We should apply this limit at every level
of a written document ... By chunking information the
writer improves the reader's comprehension and access
and retrieval speed. Since readers can at best retain no
more than 5 to 9 pieces of information in short-term
memory, they comprehend material that has been
'chunked' more quickly and more completely." (Horn
1992, p. 3-A-2)

Horn goes on to say that we should:

"... apply the Chunking Principle to:

– sentences
– blocks [that is, a group of 7 ± 2 sentences about
  a common topic, or a list or a table]
– maps [that is, a group of 7 ± 2 blocks]
– sections [that is, a group of 7 ± 2 maps] and
– chapters [that is, a group of 7 ± 2 sections]."

Also:

"Remember that the Chunking Principle advises 7 ± 2
items in a list." (Horn 1992, p. 10-5)

And:

"When your sentence is more than twenty words long,
consider dividing it."(Horn 1992, p. 12-3)

"[A] sentence [should] never be more than 30 words."
(Horn 1992, p. 12-2)

## What is a piece of information?

So writing should be grouped into units of information—blocks, maps, sections, etc.—and a "manageable unit of information is one consisting of no more than nine pieces of information". But what is a *piece* of information?

Imagine a map with seven blocks and with seven sentences in each block. This would fall within Horn's $7 \pm 2$ limit. Such a map would have $7^2$ (or 49) sentences. So if a unit of information—in this case a map—cannot have more than nine pieces of information in it, then obviously a sentence is not what Horn means by a *piece* of information.

Now if a map can have no more than nine *blocks* and no more than nine *pieces of information*, it would seem that a block and a piece of information are considered the same thing. However, a block is also a "unit of information" (Horn 1992, p. 3-B-1), so if a unit of information is composed of pieces of information, a block cannot *exclusively* be a piece of information. Otherwise a block would always have just one element in it: one piece of information.

So it seems that for Horn what constitutes a *piece* of information varies from one unit of information to another. At the level of chapter, a piece of information is a section; at the level of a section, a piece of information is a map; at the level of a map a piece of information is a block and at the level of a block a piece of information is a sentence. But what is a piece of information at the level of a sentence?

This is an important consideration. The sentence is the fundamental, indeed necessary, building block in every document. Horn accepts that this is so: "The first and basic *unit of information* is the sentence." (Horn 1992, 12-3. Emphasis added.) But he also says that "Writers should group information into small, manageable units [and a] manageable unit of information is one consisting of no more than nine pieces of information" (Horn 1992, 3-A-2). So we would expect Horn to tell us what a piece of information *in a sentence* might look like. If the sentence is the first and basic unit of information, then the entire Information Mapping edifice balances or topples on the answer to this very question. Alas, Horn does not provide an answer. And for all its apparent importance, *sentence* does not even get a mention in the index of Horn's book. Neither does *piece of information* nor *information, piece of*. We have to work it out for ourselves.

A piece of information at the level of a sentence cannot be the same as a character or a word, for Horn allows up to 30 words per sentence, as we've noted. A thirty-word sentence would then have at least thirty pieces of information in it, well above the specified limit of nine. Anyway, words like *the, a, an* and the like couldn't possibly be pieces of information. Perhaps no word on its own could be a piece of information.

After words, the next level of granularity in a sentence is a *phrase*. A phrase is a string of words that, although potentially meaningful when combined with other words, does not have a subject (a thing singled out for discussion) or a predicate (something said about whatever is singled out for discussion). Put another way, "a phrase is a group of words that acts together as a unit within a sentence [but which] can't stand on [its] own and make a sensible message" (Peters 1989, p. 339). If a phrase has no subject nor can stand on its own to make a sensible message, then obviously it cannot be a piece of information.

The next level of granularity is the *clause* (that is, a string of words with a subject and a predicate). In fact, the clause is as far as we can go. After a clause, we have a sentence, and it would make no sense to say that the *fundamental* piece of information in a sentence is the sentence itself while at the same time allowing that a sentence, being a unit of information, can have up to *nine* pieces of information.

And now it should be clear why Horn did not extend his 7 ± 2 limit to the pieces of information that make up a sentence. The prospect of reading a nine-clause sentence would repulse most readers. Even a five-clause sentence would be indigestible to many readers, even those familiar with the topic. (A sentence composed of five independent clauses is equivalent to *five separate sentences* glued together with conjunctions and punctuation.) And this, no doubt, is why Horn opted instead for a word limit, not a pieces-of-information limit, on sentences: "an average of 20 words and never more than 30" (Horn 1992, p. 12-2).

The sentence is not the only unit of information that Horn excludes from his 7 ± 2 limit. If a chunk as large as a chapter can be a unit of information, there seems no reason why a book or even a large report should not also be considered a unit of information. (It is just a collection of chapters as a chapter is a collection of sections.) But Horn nowhere limits the number of chapters in a book or report.

## Is there science behind Information Mapping?

Horn repeatedly states that his method is based on research. Here are just two of the many mentions:

> "Research suggests that people can best process and remember no more than seven plus or minus two pieces, or units, of information at one time." (Horn 1992, p. 3-A-2)

> "The ... method implements research-based findings on how individuals process and understand information most efficiently." (Horn 1992, p. 2-3)

Despite being repeated a number of times in the book, no references to research to back up any particular claim are provided in footnotes or endnotes. There is a substantial "core bibliography" which, we are told, provides "complete citations for the research base on which the Information Mapping methodology was built" (Horn 1992, p. B-1) and yet this bibliography includes many references of dubious relevance to Information Mapping. As one reads through it, a sense of quantity out-doing quality quickly intrudes. For example, there are references to generalist books such as *Statistical Methods* and *The Structure of Science*. How could such books be *specifically* relevant to Information Mapping? Including such general texts is about as silly as, say, Einstein, tendering a textbook on calculus in support of his general theory of relativity.

One also finds references in Horn's bibliography that seem hardly related to scholarly research of any kind, such as Whitlock's 1972 interview with Horn. Can a mere interview be classified as core research? There is also a reference to a 1945 article published in *Atlantic Monthly*, more a magazine than a research journal. Further, there are no immediately recognisable references to back up any specific Information Mapping principle, such as the chunking rule.

A fifteen-page bibliography might look impressive and suggest that much scientific rigor has gone into the recommendations in the book. But if the bibliography provides no help to the interested reader who wants to check how a recommendation is supported, it fails its very purpose. An author who makes a claim that is said to be based on research and yet provides no reference to that research is akin to the scientist who reports on some research but fails to explain to fellow scientists how the experiment was done and how it can be replicated. To expect the interested reader to read every item in a 15-page bibliography to verify that the accompanying text is soundly based on research is utterly unreasonable. Further, an "unpublished draft [of a] manuscript for [a] 1967 course at Harvard" along with three "unpublished proposal[s to the] US Air Force" are listed in *The Information Mapping Method: 30 Years of Research*, a 1999 publication of Information Mapping, Inc., the company Horn established to promote the Information Mapping method.[1] Can unpublished course notes and commercial proposals really constitute research? Quantity over quality again?

## The overlooked Miller

The American psychologist George Miller is widely credited with pointing out the potential significance of 7 ± 2 as a limit to various aspects of human cognition. In research conducted in

1. Available from http://www.infomap.com. Viewed 21 January 2011.

the early 1990s, Miller's seminal paper—"The Magical Number Seven, Plus or Minus Two", referred to earlier—shared with two others the honour of being the most cited paper in "24 well-known introductory psychology texts [that] fully covered the field of psychology" (Gorenflo & McConnell 1991, p.8f)[2] Horn tells us that he began developing Information Mapping in 1967, with the first version of his book on the subject published in 1976. Given the excitement generated by Miller's paper, it seems odd that Horn does not include the paper in his own bibliography. Let's repeat that: Miller set the 7 ± 2 debate in train, his paper is much discussed and cited (and still is), twenty years later Horn expounds on the importance of presenting writing in chunks of 7 ± 2 and tells us that this is research-based—*and yet neglects to mention Miller's paper*. That is nothing less than odd. Just as odd is the fact that when Miller's and Horn's bibliographies are compared, not a single reference can be found in one that is also in the other.

The failure to acknowledge Miller, and the sources Miller quotes, becomes even more curious when we read, on the Information Mapping website, that:

> "The chunking limit is a guideline, *based on George A. Miller's 1956 research*".[3]

Can we assume that this is Horn's view too? Well according to Horn's own page on the Stanford University website, he has been chairman of Information Mapping, Inc. since 1987.[4] This is the company he established twenty years earlier to promote the Information Mapping method. The website of Information Mapping, Inc. is http://www.infomap.com, the very site from which the quote above was taken. Perhaps it is not stretching it too far to think that the chairman of a company would endorse

---

2.  Miller's paper was cited in 22 of the 24 texts surveyed.
3.  See http://www.infomap.com/index.cfm/themethod/Mapping_FAQs (viewed 23 January 2011). Emphasis added. One might expect that a bibliography that includes a general textbook on statistics would include the very paper on which Information Mapping is based.
4.  See http://www.stanford.edu/~rhorn/a/site/HornCV.html. Viewed on 13 January 2011.

the claims publicly made about the chief product of that company—especially when the product is the creation of the chairman. Thus it is not unreasonable to think that Horn did, and still does, think that George Miller's research supports the Information Mapping method. But does it?

## Miller's magic number

For a start, let's make one thing clear. Miller did not do the research described in his justly famous 1956 paper about the limits of human cognition. He simply reported the results that other experimenters had published and tried to make sense of them, including the apparent fact that $7 \pm 2$ kept cropping up in various, unrelated studies as a limit to cognition, something he suggested, in the final paragraph of his paper, might be nothing more than "a pernicious Pythagorean coincidence" (Miller 1956, p. 96) So to say that Information Mapping is "based on George A. Miller's 1956 *research*" is a little misleading. If it is based on Miller's 1956 *paper*—as opposed to any of Miller's own research—then it must be based on the research of those who Miller quoted. But, as noted in the last section, not one of the researchers Miller quoted is mentioned in Horn's bibliography. Let's put this down to sloppy citing on Horn's part and assume that he meant to say that he based Information Mapping on the research that prompted Miller to think that $7 \pm 2$ was somehow significant in cognitive psychology.

Miller's paper covers three distinct topics, all in some way related to memory. Since the Information Mapping website quoted above does not mention which topic or topics in Miller's paper form the basis of the Information Mapping method, let's look at each one.

## The span of absolute judgment

Miller first considered a number of experiments in which subjects were asked to make absolute judgments about such stimuli as frequencies, loudness, saltiness, the size of rectangles, and the like. The objective was to determine how well humans can distinguish between differing levels of intensity of particular stimuli. All these experiments followed a similar pattern: subjects were exposed to a random sequence of varying stimuli after each stimuli had been given an identifying number. The subjects were then asked to repeat the numbers in the same order as the stimuli were randomly given. For example, a frequency of 100 Hz might be assigned the number 2, 6 000 Hz assigned the number 9, 8 000 Hz assigned the number 3, and so on. When exposed to tones at frequencies, say, of 6 000 Hz, 8 000 Hz and 100 Hz in that order, subjects were tested to see if they correctly responded 2, 9 and 3 respectively. The number of frequencies (or whatever stimuli was being used) was gradually increased until the number of correct responses dropped to zero.

In summarising an experiment conducted by Irwin Pollack in 1952 using tones of varying frequency as stimuli, Miller writes:

> "When only two or three tones were used the listeners never confused them. With four different tones confusions were quite rare, but with five or more tones confusions were frequent. With fourteen different tones listeners made many mistakes ... The result means that we cannot pick more than six different pitches that the listener will never confuse. Or, stated slightly differently, no matter how many alternative tones we ask him to judge, the best we can expect him to do is assign them to about six different categories." (Miller 1956, p. 83f)

This is a misrepresentation of Pollack's result. Pollack concluded that subjects could assign tones to only *five* categories:

> "... an informational transfer of approximately 2.3 bits is the maximum obtained. This is equivalent to perfect identification among only about 5 tones." (Pollack 1952, p. 748)[5]

Even so, the number of tones Miller quotes is 6, not 7—his magic number. Where, then, did Miller get his magic number? He got it by looking at a number of similar experiments—not just Pollack's—and taking the *average* span of immediate judgment (what he also calls the *channel capacity*, or the number of correctly identifiable *categories*, in Pollack's language):

> "... the channel capacities measured ranged from 1.6 bits for curvature to 3.9 bits for positions in an interval. Although there is no question that the differences among the variables are real and meaningful, the more impressive fact to me is their considerable similarity. If I take the best estimates I can get of the channel capacities for all the stimulus variables I have mentioned, the *mean* is 2.6 bits and the standard deviation is only 0.6 bits. In terms of distinguishable alternatives, this mean corresponds to about 6.5 categories, one standard deviation includes from 4 to 10 categories, and the total range is from 3 to 15 categories." (Miller 1956, p. 86. Emphasis added.)

Note that the channel capacities across a number of like experiments ranged from 1.6 bits (or 3 categories, when judging the curvature of lines) to 3.9 bits (or 15 categories when judging positions along a linear interval). Miller took what he thought was the mean value—2.6 bits of information—equated that to 6.5 categories, and then rounded it up to 7.

On the face of it, this looks like sloppy arithmetic. If the mean is 2.6 bits of information, it is much closer to 6 categories than 6.5, since $2^{2.6} = 6.06$. To get a channel capacity of 6.5, the average number of bits would have to be 2.7 (as $\log_2 6.5 = 2.7$). To get Miller's magical number 7, the average number of bits would have to be 2.8 (that is, $\log_2 7$). In Miller's defence, however, he was looking for a *limit* or *ceiling* on absolute

---

5.   Pollack's 2.3 bits of information is equivalent to $2^{2.3}$ correctly identifiable tones, which equates to 4.925. In general, the number of categories correctly judged (which Miller calls *channel capacity* and *span of absolute judgment*) is equal to $2^b$, where $b$ is the number of bits of information presented by the stimuli as a whole. Thus the number of bits associated with a channel capacity of, say, $c$ is $\log_2 c$. For our purpose, we are only interested in channel capacity, not bits of information.

judgment and this limit obviously has to be a whole number. In that case, rounding up rather than down makes sense.

Let's ignore Miller's arithmetic and concentrate instead on the wide range of channel capacities observed for various stimuli. Miller was surprised that the range wasn't greater, but if channel capacity—that is, the span of absolute judgment—is relevant to writing, there is a world of difference between advising writers to limit, say, the number sentences in a block to 3 rather than 15. If Horn relied on Miller's consideration of the observed *range* of channel capacities, then he needs to explain why judging meaning (or whatever) is more akin to judging the frequencies of tones (where the channel capacity is 5, given Pollack's research) than judging positions along a linear interval (where the channel capacity is 15). To opt without reason for 7 is hardly scientific, given that Miller did not report any experiments on the absolute span of judgment when written material was the stimuli. Indeed, for all that Miller tells us, that span could fall outside the observed 3–15 range.

Anyway, the relevance of any span of absolute judgment to effective written communication is tenuous. Readers do not have to make the sort of comparative assessments of stimuli Miller considered in order to understand what they are reading. That is, understanding a piece of written material simply does not involve recalling the relative intensities of some semantic stimulus or other. There is, in other words, no parallel between naming tonal frequencies according to a provided legend and unravelling the meaning of a map, block or sentence. And, as we'll see a little later, Miller said that himself.

## The span of attention

The next part of Miller's paper—entitled *Subitizing*—has to do with "the discrimination of number". The experiments that Miller reports have long precedents. In the nineteenth century, the Irish mathematician William Hamilton and the English economist Stanley Jevons independently showed that if subjects were briefly shown marbles (or stones) in a box, they could easily remember the number they had seen up to about seven, after which accurate recall fell away. Miller quotes more recent experiments that give the same result, in this case, dots flashed on a screen for 0.2 seconds at a time. An accurate count drops away dramatically after seven dots.

Could this be what Horn was basing Information Mapping on? It seems unlikely. No writer writes with the expectation that readers will only see what they have written for 0.2 seconds at a time (or two minutes, for that matter).

## Span of immediate memory

The last part of Miller's paper is taken up with what he calls the *span of immediate memory* (also known as the capacity of our short-term memory). This may be what Horn based his method on given his repeated appeal to the limit of our short-term memory:

> "Since readers can at best retain no more than 5 to 9 pieces of information *in short-term memory*, they comp-rehend material that has been 'chunked' more quickly and more completely." (Horn 1992, p.3-A-2. Emphasis added.)

In the two experiments that Miller cites, subjects were given several stimuli in succession and then asked to immediately recall them. The types of stimuli included binary digits, decimal digits, letters of the alphabet, letters plus decimal digits and monosyllabic words. The results:

> "With binary items the span [of immediate memory] is about nine [and it] drops to about five with mono-syllabic English words." (Miller, 1956, p. 92)

A span of 5 to 9 looks suspiciously like Horn's magic range of 7 ± 2. But just how relevant to comprehension is a measure of our ability to *immediately* recall a list of digits, words or whatever? Take one of Horn's maps, for example. A map can have 7 ± 2 blocks and each block can have 7 ± 2 sentences (or a list of 7± 2 items). Now consider the simplest map possible while keeping within the 7 ± 2 range: a map with five blocks each with a list of five items. Now try this experiment: on each of five cards write a list with five items in it. If you like, make it easier for yourself to remember the items by including only like items in each list (say the names of birds in one list, cities in another and so on). As you complete a list, turn the card over so that you cannot see what you have just written. When you have finished writing all five lists, try to recall, without looking at the cards, every item in all five lists in the order in which you wrote them. What are the chances of you correctly recalling all items? About the same as getting a money-back guarantee from a palm reader. What about recalling all the items in any order?[6] Only marginally better. You will probably recall the first few items and last few—what psychologists call the *primacy* and *recency* effects—but struggle with those in the middle. But if Miller's recall research is relevant to comprehension at the level Horn envisages, you should be able to correctly recall all 25 items, since they were in just five chunks (that is, five blocks). But if you can't correctly recall the items, have you comprehended the material?

It might be retorted that a test of memory is not a test of comprehension, and that is true—in some cases. I might be able to recall the sentence "The work done was five ergs" but have no comprehension of what it means if I haven't studied physics. But in the case where you have listed the words to be recalled yourself, it is difficult to see what comprehension could be other than recall. Ability to recall at a later date, perhaps? Well that's hardly likely to be better than your ability to recall immediately after listing the items. Anyway, we are now straying far from Miller's paper, which we are told is the foundation for the Information Mapping method. Miller never examined delayed recall. Nor did he examine comprehension.

We poorly remember five lists of five items (and even shorter material) because, by definition, short-term memory is short:

> "Forgetting over intervals measured in seconds was found." (Peterson & Peterson 1959, p. 198)[7]

---

6.  It's not clear from Miller's paper whether the tests discussed were of serial recall (recall in the order given) or free recall (recall in any order).

"... 30 seconds is ample time for forgetting to occur".
(Baddeley1987, p. 10)

This makes Horn's view that the limited capacity of our short-term memory should compel writers to apply the 7 ± 2 chunking rule at the *molecular* level—at the level of blocks, maps and sections—decidedly odd. By the time I've read even one seven-chunk block in, say, a map, 30 seconds is likely to have passed. If there are seven blocks in the map, many minutes may have passed by the time I have read them all. That is, as Baddeley says, "ample time for forgetting to occur". Further, a map could have over 50 items in it: 7 blocks each with a list of 7 items, plus all the headings. Given Miller's limit, my short-term memory will have been flushed out many times by the time I reached the end of that map. So what is the relevance of limiting a map to seven blocks? If 7 ± 2 applies only to *short-term* memory—and that is all that Miller said—why must it apply to chunks of material that could never be accommodated in a single frame of short-term memory?

The capacity of our short-term memory might well be relevant to our ability to comprehend material at the *atomic* level of a text—the clause or sentence—but at the *molecular* level—the level of blocks, maps, sections and chapters—its relevance is doubtful. If this is what Horn is claiming, then the onus is on him to advance some supporting research.

## Flawed reading assumptions

Even if Miller had been right in thinking that the span of immediate memory is limited to 7 ± 2, the way this span was established—by testing *immediate* recall—and the span itself are quite irrelevant to the way people read. Suppose I want to know how to use time-shift on a personal video recorder and consult the accompanying user guide for instructions. Suppose, too, that I encounter a seven-step procedure (indeed, the

---

7. The experiments reported in this paper suggest that after about four seconds, correct recall drops to about 50%, and to zero after 18 seconds. See figure 3 on page 195 of the paper.

number of steps is irrelevant for this argument). Am I meant to read each step and then recall them all before I can successfully complete the procedure? Of course not. Further, in a recall experiment such as those Miller considered, the subject cannot ask for the items to be repeated; in the real world a reader can go back and read an earlier step in a procedure, if they really needed to. The relevance of Miller's work is looking shakier and shakier.

Further, hierarchical chunking — from block to chapter — is unlikely to help most readers of many types of documents. Consider the types Horn was primarily concerned with: procedures and policies. Most readers don't read whole chapters in user manuals and sets of work instructions, the types of documents where procedures and policies are prevalent. They dip into such documents when they want to learn (or be reminded of) how to do something in particular. They want to activate time-shift on their personal video recorder and the steps are not obvious. They might then scan the contents pages or index, or electronically search, for the topic of interest and then read *just that topic*. Perhaps in the effective life of a personal video recorder, the owner might consult the user guide a dozen times but never read it right through. Much to the chagrin of technical writers, a user guide is primarily consulted as a last resort: when the product it describes does not work as the user expected it to, or when the information needed cannot be got by asking someone else. And when it is, only a small part of it is consulted at any one time.

In Information Mapping, there is a special type of map for procedures and work instructions: the procedure map. If most readers of instructional materials only dip into the materials to learn how to do a particular task, on each reading they are likely to be reading only one, perhaps two, maps. In which case they will not need any special guidance that might be given by the $7 \pm 2$ structure of a *section*. (Recall that a section is a group of $7 \pm 2$ maps.) Nor will they need any special guidance that might be given by the $7 \pm 2$ structure of a *chapter*. In other words, readers of procedures and work instructions are more interested in the trees than the wood. Indeed, most won't even see the wood.

Let's put this another way: will I *understand*, say, a procedure more quickly or more thoroughly if the chapter of which it is only a small part has seven rather than, say, ten sections? Unlikely. The *molecular* structure of the chapter will simply be unnoticed (and how can something I don't notice influence my degree of comprehension). Will I *find* that procedure more quickly with a seven-section structure than a ten-section structure? Hardly, given that I, like most readers, will go to the index or table of contents for help in finding a procedure. (And even if there were no index or table of contents, molecular chunking is unlikely to improve the speed with which topics are located. Indeed it may *impede* that speed, given that complex topics that should logically be kept together—and which readers would expect to find together— might have been split across chapters, and for no other reason than to avoid breaching Horn's chunking limit.)

All this puts paid to Horn's claim that:

> "By chunking information the writer improves the reader's comprehension and access and retrieval speed."
> (Horn 1992, p. 3-A-2)

The *molecular* structure of a document has no bearing at all on my ability to understand any *atomic* part of that document; nor does it necessarily improve how quickly I can find a particular part of that document.

Even if readers did notice the molecular structure of a document, how can the number of items in any particular structure or sub-structure interfere with one's ability to understand material at the atomic level? Suppose that two people of otherwise equal intelligence, education and experience are asked to read and comprehend a single paragraph (the same paragraph in both cases). Let's call them *A* and *B*. In *A*'s case, the paragraph to be read is embedded in a group of seven paragraphs. In *B*'s case, the paragraph to be read is embedded in a group of 10 paragraphs. *A* and *B* are both asked to count the number of paragraphs presented to them before reading, and answering questions about, just the specified paragraph. Can we conclude that $A$—with fewer paragraphs in the material presented—will better comprehend the paragraph? In other words, will $A$ score higher than $B$ on the comprehension questions? It would be surprising if any evidence could be found to support that view. Certainly Miller didn't do such an experiment, and thus the claim that Information Mapping is based on Miller's research is looking even slimmer still. Indeed, the *prima facie* silliness of the idea that the super-structure of a document affects comprehension of a sub-structure puts the onus on information mappers to advance some plausible evidence in support of the claim that the limited capacity of our *short-term* memory helps us comprehend anything other than atomic information (such as a clause or sentence).

## Is Miller relevant but memory not?

Horn's main point is that "since readers can at best retain no more than 5 to 9 pieces of information in short-term memory [writers are compelled to apply a chunking limit of 7 ± 2] at every level of a written document" (see page 112). Short-term memory, it seems, is important. It is *the* determining factor. We also noted that the website of the Information Mapping company of which Horn is the chairman states that "the chunking limit is a guideline, based on George A. Miller's 1956 research" (see page 118). As we've just seen, short-term memory is the very stuff of Miller's paper, especially its role in judgment, attention and recall. What, then, are we to make of the rest of the reference to Miller on the Information Mapping website:

> "The chunking limit is a guideline, based on George A. Miller's 1956 research, for creating information that people have to memorize. Documents do not have to be 'memorized', but maintaining these chunking limits aids in a reader's ability to process information."[8]

---

8.  See http://www.infomap.com/index.cfm/themethod/Mapping_FAQs (viewed 23 January 2011).

This is puzzling. If readers don't need to memorise material they read—which is true—then why must that material be limited to chunks that do not exceed the capacity of short-term memory? And why say that the chunking limit is based on Miller's research when Miller's research says nothing about what might aid a reader's ability to process information *other than information that has to be immediately recalled*. If we are not discussing information that needs to be immediately recalled—which we can't be if memorisation is not an issue— then we can't be basing our research on Miller. So where is the research that shows that even though memorisation is not required of readers, writers must "at every level of a written document" chunk their material in line with the constraints imposed by the capacity of short-term memory? Claiming Miller as an authority is looking a touch like fabrication.

But the very same web page that tells us that "Documents do not have to be memorized" also tells us that:

> "Chunking ... involves making the information digestible either for memorization or comprehension."[9]

So perhaps all that the first quote meant was that we don't have to memorise *whole* documents in order to understand their contents. That's too obvious to warrant discussion. But what about parts of a document? It's worth repeating here that Miller was concerned solely with *immediate* recall. Does anyone ever attempt to—or even need to—memorise a chapter, section, or even a map for *immediate* recall? Of course not. Does anyone even attempt to memorise a chapter, section, or map for *later* recall? Again, of course not. I might prefer to remember the main points in, say, a map so that I don't have to resort to the user guide in the future. But if my attempts at memorisation are not tied to the limits of my short-term memory—if I have time, that is, to rehearse and to mnemonically code what I want to remember—then Miller's magic number, pertaining as it does to *immediate recall*, is utterly irrelevant. And if the concern is to minimise the effort

---

9.   ibid.

involved in rehearsing and mnemonically coding what I want to remember, then perhaps we should be applying the *minimum* chunk limit possible before recall errors occur: three (as noted in the next section). Miller was, you may recall, talking about the *limits of perfect recall*.

A parallel should reveal the absurdity of Horn's position. A listener does not need to memorise a melody, section or movement of a piece of music in order to appreciate it. (Otherwise we would never be enthralled by a piece of music we had never heard before.) Now let us suppose, for the sake of argument, that humans can only accurately hum 7 notes of any melody they hear for the first time when asked to immediately recall it. Does it then follow that composers should restrict to seven the number of notes in a melody, the number of melodies in a section, the number of sections in a movement, and so on? Again: of course not.

## Research after Miller

In a 98-page paper published in 2000, American psychologist Nelson Cowan reviewed the data then available on the limits of short-term memory. He noted that researchers post-Miller had found that short-term memory is limited to between three and five chunks. In summarising the masses of data he reviewed, Cowan concluded that there is a:

> "single, central capacity limit averaging about four chunks ..." (Cowan 2000, p. 87)

He also noted that:

> "[Miller's magic] number was meant more as a rough estimate and a rhetorical device than as a real capacity limit."

Perhaps the Information Mapping fraternity might like to update their research instead of relying on Miller's survey of a handful of experiments conducted more than 50 years ago. If the span of short-term memory is only 4 ± 1 and if, as Horn claims, the span of short-term memory sets the chunking limit, then

The pseudoscience of Information Mapping

Information Mapping needs to be radically updated to bring it into line with current knowledge in cognitive psychology.

However, Cowan's reported limit is a limit on the recall of *unrelated* items (as was the limit reported by Miller). But when psychologists look at strings of *related* words—such as words in a sentence—the span of immediate memory is much greater:

> "Immediate memory for sentential material is typically substantially greater than span for unrelated words ... Baddeley *et al.* ... found spans of around five for unrelated words [in line with what Cowan reported] and 15 [words] for sentences." (Baddeley 2007, p. 143)

So if the span of short-term memory determines the chunking limit—as Horn contends—and if the span of short-term memory *for the sort of material that writers primarily present to readers* (namely sentences) is 15, perhaps the chunking limit should be raised to 15. Either way, Information Mapping has been left behind by research more recent than Miller's. A limit of 7 ± 2 is yesterday's guesstimate. Today it is 4 ± 1 for unrelated items and 15 words for sentences, the very entities that Horn claimed are the "first and basic unit of information" (see page 114).[10]

## What George Miller might say about Information Mapping?

Well-known author and freelance editor Mark Halpern wrote to George Miller in the mid-1990s when confronted with a workplace edict to limit the items in a list and the steps in a procedure to 7 ± 2. Halpern knew that Miller's name was associated with research on the limits of cognitive processing and that Miller had publicly complained about the unfounded conclusions some had drawn from his research. Miller replied to Halpern detailing one of those unfounded conclusions. In

---

10. Note that Horn's omission of sentences from his 7 ± 2 chunking rule (see page 115) is not relevant here. We are only considering the raw capacity of short-term memory, which Horn uses to justify the chunking of blocks, maps, sections and chapters.

the 1970s, some local authorities had passed by-laws restrict-ing the number of items that could be displayed on a billboard to 7 ± 2, using Miller's research to justify the laws. (It turned out that a group of landscape architects, funded by the big motel chains, had lobbied the authorities to introduce the law.) In his reply to Halpern, Miller said:

> "the point was that 7 was a limit for the discrimination of unidimensional stimuli (pitches, loudness, brightness, etc.) and also a limit for immediate recall, *neither of which has anything to do with a person's capacity to comprehend printed text.*"[11]

To sum up:

- Information Mapping is not based on Miller's research nor on any research that Miller quoted. To claim that it is ignores, or misunderstands, what Miller actually wrote.
- Even so, Miller's research has been superseded by more recent studies.
- Information Mapping is at odds with the way people read texts. The molecular structure of a text is rarely if ever noticed by readers (whether seeking spur-of-the-moment information or in the exceptionally rare case where a reader reads the text all the way through).
- Even if readers did notice the molecular structure of a text, no plausible evidence has been adduced to support the claim that the limited capacity of our *short-term* memory restricts our ability to comprehend anything other than atomic information (such as a clause or sentence). At the molecular level of chapters, sections, maps and blocks, the capacity of our short-term memory appears entirely irrelevant.

---

11. The entire thread between Halpern and Miller can be read at http://members.shaw.ca/philip.sharman/miller.txt. Emphasis added. Viewed 14 January 2011. In an email to me on 8 February 2011, Halpern confirmed the accuracy of his reported exchange with Miller. An email to Professor Miller at his last-known Princeton University address bounced.

To conclude: a document-wide chunking limit of 7 ± 2 is not science. It is pure bunk.

## Part 2: How can size matter?

In the first part of this paper, we dissected the claim by Robert Horn that research by American psychologist George Miller on the limitations of short-term memory shows that we should present information to readers in chunks of no more than 7 ± 2 sub-chunks. To repeat Horn's prescription to technical writers:

> "Writers ... should apply this [7 ± 2] limit at every level of a written document ... By chunking information the writer improves the reader's comprehension ... since readers can at best retain no more than 5 to 9 pieces of information in short-term memory ..."[12]

We found that Miller's research on the limitations of short-term memory does not support Horn's claim, and Miller himself was quoted as saying so.

The fact that Horn erred in basing his chunking principle on the limited capacity of our short-term memory does not on its own disprove that some chunking limit is necessary for comprehension. Perhaps, then, the chunking principle at the heart of Information Mapping is still appropriate even if the work of Miller cannot be adduced in support of it.

Let's assume, for the sake of argument, that there is some chunking limit. To what chunks in a document might it sensibly apply? Horn contends that it should apply to *every* structural component in a document: blocks, maps, sections and chapters.

First note that there is nothing in Horn's claim that "chunking ... improves the reader's comprehension"[13] to suggest that he is using the word *comprehension* other than in its common-or-garden dictionary meaning:

---

12. "The chunking limit is a guideline, based on George A. Miller's 1956 research". See http://www.infomap.com/index.cfm/themethod/Mapping_FAQs (viewed 23 January 2011).
13. ibid.

"**comprehension** *noun* 1. the act or fact of comprehending"

"**comprehend** *verb (t)* 1. to understand the meaning or nature of"[14]

Thus Horn's claim is that chunking improves a reader's ability to *understand* what they are reading. Analysed this way should make it blindingly clear that the relationship between chunking and *comprehension-as-understanding* is a tenuous one indeed. If such a relationship exists at all, then it plainly cannot apply to *all* types of writing. A novel, for instance, is largely unchunked. It will have paragraphs (that is, blocks) and it may have chapters. But it won't have maps and very few novels have sections. And of the chunks a novel does have, no limit is adhered to by the author. Does this lack of rigorous chunking in the form required by Information Mapping mean that we do not understand what we read in novels? To retort that chunking *would* improve comprehension is to imply that without chunking every reader must necessarily fail to understand some parts of a novel. Or, to put it another way, no-one has ever fully understood a novel. That's a tough claim to support.

Comprehension is often measured by a reader's ability to *recall* the salient facts in what they have read. Perhaps what Information Mapping has in mind in claiming that "in chunking information the writer improves the reader's comprehension" is that chunking improves a reader's ability to correctly recall what they have read. The bigger the chunk they have to read, the more difficult it is for a reader to recall the salient facts in the chunk. This matches the folk wisdom that the longer the paragraph, the more difficult it is to understand:

> "For general purposes, paragraphs from 3 to 8 sentences long are a suitable size for developing discussion, and some publishers recommend an upper limit of 5/6 sentences." (Peters 2007, p. 595)

> "Paragraphs should be kept short wherever possible." (ISO/IEC 26514 2008, p. 94)

---

14. Macquarie Dictionary, http://www.macquariedictionary.com.au. Viewed 11 May 2011.

Note, for a start, that if we were asked to read a paragraph *for recall*, we would read it differently than if asked just to read it. (By *recall* here I mean the ability to repeat the main points in what one has read, not the ability to accurately regurgitate what one has read in the order that it was read.) To read a text for recall is to study that text, and to study a text is not the same as reading a text. Our reading strategies are quite different. Studying involves mental rehearsal, repetition and summarisation, and possibly note-taking. Through such strategies we improve our chances of correct recall. But these strategies remain idle when we read a newspaper, novel, article or report. Here we are happy to absorb each sentence as it is presented to us free of the mental gymnastics required in, say, preparing for an examination. A student might take hours to study a section in a textbook that the casual reader might read in 30 minutes.

In other words, we must be careful not to confuse *understanding* with *knowing*. I can understand how to replace the drive belt on a robot by understanding the steps in the procedure and replacing the belt. But if I wasn't expecting to be *tested* on my knowledge of the procedure, I might subsequently fail a closed-book exam that asked me to outline the steps required to replace the belt. To repeat: our reading styles when we are *studying* differ markedly from our reading styles when we are just reading for immediate throw-away facts. We might call the two styles of reading *studious reading* and *transient reading*. Students mostly engage in studious reading—at least when studying—while the rest of us engage in transient reading (at least when reading informational texts). We are reading simply to gain immediate information: for one-off understanding or one-off use. And a reader can understand a sentence without necessarily being able to recall it, or its informational content, hours, days or weeks later. (Did you understand the sentences in this paragraph? Make a note to yourself to try to recall its salient facts tomorrow.)

For the sake of argument, let's suppose that Information Mapping's chunking limit was found to be necessary to maximise *recall* (a point that has not been proved, incidentally). If so, what reason might there be to apply the chunking limit to "any business writing task and any type of document" (Horn 1992, p. 2-1) and "at every level of a written document" (Horn 1992, p. 3-A-2). Why force writers *at all times* to adopt a methodology supposedly best suited for study and recall when most readers don't read documents for study and recall? Most of us don't want to clutter our minds with one-off facts and throw-away information, especially information we know we can readily access again if we need to (by, for example, rereading a procedure or report). Does any one *study* a procedure? No. We read through it step by step (or perhaps just read those steps that we are unsure of). Likewise, we don't *study* a policy document, annual report or business memorandum. We simply read it in the hope of understanding what it is we are reading but without the burden of needing to recall what we read. In a word: if most readers engage in transient reading, why force authors to write in a way that will assist studious reading (assuming that that is the case)?

Since most readers of user guides and policy manuals—the sort of documents Horn is primarily concerned with—adopt transient reading, how might they approach that reading? From a chunk-awareness approach, there are only six logical possibilities:

1. read some part of it (a section, map or block) without first paying attention to the number of parent or grandparent chunks
2. read some part of it (a section, map or block) after paying attention to the number of parent or grandparent chunks
3. read some part of it without first paying any attention to the number of sibling chunks
4. read some part of it after paying attention to the number of sibling chunks
5. read some part of it (a chapter, section or map) without first paying any attention to the number of child or grandchild chunks

6. read some part of it (a chapter, section or map) after paying attention to the number of child or grandchild chunks.

In other words, readers look (or don't) at topics higher up the topic hierarchy, they look (or don't) at topics at the same level in the topic hierarchy, or they look (or don't) at topics lower down the topic hierarchy.

Take scenario 1: reading without first noticing any higher-order structure. This covers the vast majority of cases. Most readers will go to the index or table of contents for help in finding the chunk—the block, map or section—they want to read, go to that chunk and start reading it. The number of chunks in the *chapter* of which the block, map or section is a part will pass unnoticed. For example, if I want to learn how to set preferences in Adobe Photoshop, the index points me to page 86. I go to that page and start reading. I don't even notice that this procedure (which is a *map* in Information Mapping terminology) happens to be in chapter 1; and I don't bother— why would I?—to flick through the chapter looking at the other chunks in it. I'm interested only in learning how to set preferences. Now if I don't pay any attention to the size of the parent or grandparent chunks—the sections and the chapters in this particular example—how can the size of those chunks affect my comprehension of what I am reading?

Let's explore this scenario a little further. Suppose that A is given version 1 of a user guide and B—of equal intelligence, knowledge and experience—is given version 2. Suppose further that the only difference between the versions is that version 1 has 7 sections in chapter 1 and version 2 has 10. A and B are then asked to read, say, a map in chapter 1, a map that is identical in both versions, and to do so without first counting the number of sections in the chapter. Suppose further that the map A and B are to read is very short, something akin to the following:

1. Enter your current password.

2. Enter a new password.

3. Re-enter the new password.

4. Tap **Close**.

If a chunk-size limit did affect comprehension of a child block, we would have to say that, although this map should be very easy to understand—it has only four simple sentences in it—it would be difficult for B but not forA. Surely this is nonsensical. Simplicity is simplicity whatever may clutter around it. My understanding of "Enter your current password" seems in no way compromised by the number of unobserved sections or even maps in the grandparent chapter. In other words, a chunk-size limit is entirely irrelevant in this majority-case scenario.

If something that we do not look at in a book can affect our comprehension of something that we do look at, how might this occur? What is the mechanism of influence? A sort of spooky, quantum-like, action-at-a-distance springs to mind (much like the mechanism that entangles distant photons). Over what distances might this force field of influence spread? On the face of it, there seems no reason why it should be limited to *within* books. Might it not also extend its influence *between* books? Now there are thousands of books in my library and few, if any, would have been written with Horn's chunking limit in mind. Supposedly, then, all the messy, poorly chunked structures in all my books must each be emanating a confusing subliminal noise, together building up a cacophony so distracting as to render all comprehension impossible. Obviously, this is plain silly. So perhaps this spooky mechanism has only *intra-book* influence. Either way, we should expect a supposedly research-based documentation methodology to provide evidence that this mechanism actually exists—that B, in the example in the previous paragraph, actually has more difficulty than A in under-standing the very same map (no matter how simple the concepts in it). Until then, we'll take the advice of William of Occam and cast out this mysterious, hitherto unnoticed force field from the realm of likely influencers of comprehension.

Now consider scenario 2: reading after noticing the size of the parent or grandparent chunk. As implied when considering scenario 1, scenario 2 is unrealistic. How many of us, when we want to follow a procedure, first look at how

many *sections* there are in that chapter of the user guide? Few if
any. Likewise, how many of us, when we want to read a
section of a policy document, first count the number of
chapters in the document? Still, for the sake of argument, let's
explore the logic here. Suppose I read, in version 1 of a
manual, "Enter your current password". This is a one-sentence
block in a procedure map. Suppose further that there are 9
sections in the grandparent chapter (9 being within Horn's
chunking limit). Twelve months later, a new version of the
manual is released and now that chapter has 10 sections in it
(one more than Horn's chunking limit). This is a fact I happen
to notice. Now I go back to the procedure I read twelve months
earlier and once again encounter "Enter your current
password". Is that step now more difficult to comprehend —
*and more difficult simply because of the extra section in the chapter?*
Hardly. Or, returning to our example on page 137, will B find
the four-sentence map more difficult to comprehend than A
simply because B *notices* that version 2 of the user guide has
more sections in a chapter than allowed by Horn's chunking
limit? Again, unlikely. So a chunk-size limit is irrelevant in this
scenario too.

Consider now scenario 3: read some part of a chunk
without first paying attention to the number of sibling chunks.
This would be a fairly common reading practice. Suppose I
start reading a procedure (that is, map) without noticing that
there are 10 or more other procedures in this same section of
the user guide. For that unnoticed chunking excess to
influence my comprehension of the procedure I do read, we
need once again to introduce our spooky force field, for how,
without it, can we claim that what we don't notice influences
our ability to comprehend what we do notice.

Suppose that A is given version 1 of a user guide and B — of
equal intelligence, knowledge and experience — is given
version 2. Suppose further that the only difference between the
versions is that version 1 has 7 procedures in section 1 and
version 2 has 10. A and B are then asked to read, say, a
procedure in section 1 that hasn't changed between versions

and to do so without first counting the number of procedures in that section. Suppose further that the procedure to be read is very short, something akin the password-changing procedure discussed on page 137. If a chunk-size limit did affect comprehension of a sibling procedure, we would have to say that, although this procedure should be very easy to understand, it would be difficult for B but not forA. Our reason: an unnoticed feature of version 2—being that the sum total of like procedures in that section breaches Horn's chunking limit—somehow interferes with B's comprehension of the procedure, an interference that A does not suffer from. Surely this is nonsensical. AS we said earlier: simplicity is simplicity whatever may clutter around it.

Consider now scenario 4: read some part of a chunk after paying attention to the number of sibling chunks. Again, this is an unrealistic scenario. How many of us count, or check, the number of procedures in a section before working our way through one of them? Few if any. Suppose that I read, in version 1 of a manual, "Enter your current password" as the first step in a nine-step procedure. Suppose further that I notice that there are 8 other procedures in this section of the user guide. Twelve months later, a new version of the manual is released and there are, I notice, now 10 procedures in the same section. Will I now struggle to understand "Enter your current password", and solely because of the extra procedure in the section? Again, an affirmative answer seems absurd. A chunk-size limit at the sibling level seems irrelevant.

Consider now scenario 5: read a chunk without first counting the number of sub-chunks. This is probably the only realistic scenario, the one that most closely matches how we read. But it too has problems for an information mapper. Suppose I start reading a procedure (that is, map) without noticing that it has $n$ blocks in it (say, $n$ steps). Perhaps there are 5 steps on one page and $n - 5$ on the next, with the latter not observed until I turn the page. Can it possibly be that my comprehension of any one of the first 5 steps is somehow influenced by the number of steps on the next page? Will I better understand "Enter your current password" if there happens—unbeknown to me—to be just 2 steps on the next page rather than 20? Again, an affirmative answer would require us to postulate some spooky action-at-a-distance to explain how what we don't see influences our understanding of what we do see. And the same applies regardless of what chunk we consider. For example, my ability to comprehend a section is not compromised by the number of unnoticed sibling sections. Thus a chunk-size limit is irrelevant in this scenario too.

Finally, scenario 6: we now take notice of the number of sub-chunks. Consider the following 10-step procedure, the steps in which I count before I tackle it:

1.  Add 1 and 1 and write down the answer.
2.  Add 1 and 2 and write down the answer.
3.  Add 1 and 3 and write down the answer.
4.  Add 1 and 4 and write down the answer.
5.  Add 1 and 5 and write down the answer.
6.  Add 1 and 6 and write down the answer.
7.  Add 1 and 7 and write down the answer.
8.  Add 1 and 8 and write down the answer.
9.  Add 1 and 9 and write down the answer.

The number of steps in this procedure is within Horn's chunking limit. So the procedure should not be difficult comprehend. Remember that we are discussing transient reading here, not studious reading. I am reading this procedure to *understand* it, not to be able to later recall its details.

Suppose, now, that the following step is added:

10. Add 1 and 1 and write down the answer.

The number of steps in this procedure is now over Horn's chunking limit and thus the procedure should be difficult to comprehend. But just how difficult is 1 + 1? Indeed, how difficult is any step in this procedure?

Further, when does the difficulty arise? Do we fully comprehend all the sub-chunks up to the limit (9) and then struggle with those beyond it ("Add 1 and 1 and write down the answer")? Or does the observed breach of the limit somehow spread its tentacles throughout the entire chunk and affect our comprehension of every sub-chunk in it (even a step as simple as "Add 1 and 1 and write down the answer.")? Neither possibility has an iota of plausibility. Thus we can also discard scenario 6.

It follows from our six scenarios—scenarios that cover all possible reading practices—that comprehension does not impose any limit on the size of the chunks in a document. Comprehension may well impose a limit on the size of sentences—based on the capacity of our short-term memory[15]—but beyond that, size does not matter. Thus the chunking principle at the heart of Horn's Information Mapping methodology is fundamentally flawed.

## Part 3: Paragraph length and comprehension

Horn's chunking limit $(7 \pm 2)$ might be spurious, but some limit, surely, must apply if comprehension is not to be compromised. Or must it?

Information Mapping assumes that the sentence is the fundamental unit of information:

> "The first and basic unit of information is the sentence."(Horn 1992, p. 12-3)

Linguists would dispute this. Indeed, a moments reflection will reveal that a unit of information more fundamental than the sentence is the *clause*. A clause is a string of words with a *subject*—something that is singled out for discussion—and a *predicate*—something that is said about the subject. A sentence can be a single clause—such as *The experiment was a failure*—or an amalgamation of several clauses, as in *The cyclone has passed and not a single building is still standing*. In the first sentence, there is a single unit of information: the experiment failed; in the second there are two units of information: there was a

---

15. This is discussed in "Does sentence length matter?" starting on page 57.

cyclone and it destroyed all the buildings. Similarly, a three-clause sentence offers three units of information, as in *Since we missed the last bus and there were no taxis about, we walked home.*

Suppose that there is some sentence-based chunking limit to paragraph comprehension and suppose further that it is 9 (in line with Information Mapping's 7 ± 2, although the actual limit is not pertinent to this discussion). Suppose further that you have two paragraphs: *A* and *B*. *A* is composed of 7 sentences each of 3 clauses (making 21 units of information in total), and *B* is composed of 10 sentences each of 1 clause (making 10 units of information in total). On the face of it, we seem to have a paradox: the paragraph that meets the chunking limit (*A*) has twice as many units of information in it — and thus you would expect it to be harder to comprehend — than the paragraph that exceeds the chunking limit (*B*). How can this be so?

Of course, there may be no difficulty if we assume transient reading. But let's assume that *studious* reading is being attempted, where readers can adopt strategies of rehearsal, repetition and summarisation to remember the salient points, unencumbered by a time limit. With studious reading, it is possible for someone to recall all 21 units of information in paragraph *A*.

But in accepting this we must also conclude that the chunking limit, whatever it might be, is irrelevant in studious treading. If I am motivated to learn, there is no time limit on my study and I adopt effective study practices, there seems no limit to the number of units of information I could recall. How else does one pass an exam given the convoluted, verbose, multi-clause stew commonly served up in academic textbooks? How else do people memorise $\pi$ to a thousand and more decimal places?

So, either a chunking limit leads us to a paradox — more is easier than less to comprehend (transient reading) — or it is irrelevant (studious reading).

## What does the research show?

For the sake of argument, let's explore the idea that there must be a sentence-based chunking limit for paragraph comprehension *understood as the recall of salient points*. For a chunking limit to be useful in ensuring maximum recall, there must be a sentence-count below which recall is near enough to perfect and beyond which recall begins to deteriorate despite how studious the reader has been. That is what is meant by a *limit*. As Miller reported in experiments on the span of immediate memory—noted in the first part of this paper—the immediate recall of, say, a list of arbitrary digits is usually close to perfect up to about 7 digits and then starts to fall away. Is there a similar limit ($n$) in relation to paragraphs such that information is correctly recalled in paragraphs of up to $n$ sentences but not so in paragraphs of more than $n$ sentences despite readers engaging in studious reading? Yes, there will always be memorisation freaks who pull off dazzling feats of recall. But is the common-or-garden reader swatting for an exam likely to find their study strategies thwarted by long paragraphs?

The American psychologist Walter Kintsch has conducted a number of experiments on paragraph recall. Unlike Horn, Kintsch well understood that sentence count alone is too blunt a measure when sentences can range widely in clausal complexity. Even clauses contain individual bits of information the complexity, or number, of which might affect recall. Rather than focus on sentence number or clause number, Kintsch decided to test whether conceptual density and conceptual uniqueness influences recall. (Kintsch et. al. 1975). In one experiment, Kintsch gave subjects texts of varying conceptual densities and, after they had said they had had enough time to learn what was in the texts, they were asked to recall the propositions (or main points) made in each text. Subjects did not have to repeat verbatim what they had read; just recall, in their own words, what the salient points were in a text they had just read, and read without a time limit.

Kintsch found that subjects could recall less information from a paragraph that had more uniquely mentioned concepts than from a paragraph with fewer uniquely mentioned concepts *even if the paragraphs had the same number of sentences*. So *conceptual density*—the number of uniquely mentioned concepts—is more a determinant of successful recall than paragraph length. The greater the number of once-mentioned concepts, the more difficult it will be to accurately recall a text's main points, even when you have been given as much time as you like to study the text. Despite the repeated claim that Information Mapping is "research-based", there is no mention of Kintsch's research in Horn's book.

But what is more pertinent to our discussion are the *absolute* recall rates Kintsch discovered. Even with short paragraphs—those with just two sentences, and 21–23 words in total—recall rates as low as 58% were observed.[16] The recall rates for longer paragraphs—those with three sentences and 67–75 words in total—were lower (down to 36%), but that is not relevant to this discussion (and hardly surprising). What is relevant is that even with short paragraphs—those of just 2 sentences—recall rates were as low as 58%. So if there is a limit to paragraph length below which there is perfect recall—that is, a recall rate of 100%—that limit must be less than 2 sentences. Thus a close-to-perfect recall rate is only going to be possible with paragraphs of one sentence. In other words, if there is a chunking limit for perfect recall, it can only be one.

This would seem to contradict the claim we made at the end of the last section. When readers engage in studious reading they should be able to recall more than the contents of one sentence per paragraph. How else do we succeed at school and university? The answer lies in motivation: the more motivated we are the more likely we will adopt special learning strategies, such as summarising, rehearsal and

---

16. Kintsch et al. 1975, p. 202, table 7. The number of propositions—that is, amount of information—subjects were asked to recall, and the size of each paragraph, are given in table 1 on page 198. The size of the paragraphs is given in terms of the number of *words*. The sample texts Kintsch gives tells us the size of the paragraphs in terms of the number of *sentences*.

mnemonic encoding. Kintsch's subjects were obviously not especially motivated to recall what they were asked to read.

It should be clear now that a chunking limit based solely on sentence-number is worthless as a guide to maximising recall.

## Later research

If paragraph length understood as the number of sentences in it is a poor indicator of the likelihood of correct *recall*, might it be a better indicator of our ability to comprehend a paragraph when we are engaging in *transient* reading? We challenged this idea in the previous section, but let's assume, for the sake of argument, that this is still an open question.

How can we measure comprehension of material read transiently? Amount recalled would be a poor measure of transient understanding if those whose understanding is being measured are aware that it will be measured by the amount of material they can subsequently recall. It is only natural for subjects to want to get the best score possible on a comprehension test, and thus they will be disposed to adopt studious reading rather than transient reading as their approach. However, someone can understand what they are reading without necessarily being able to subsequently recall it. Thus transient understanding is best measured in some way other than by recall.

Perhaps aware of this Heisenberg-like uncertainty —that as soon as you try to measure *understanding* it becomes something else: *cognitive lodgement*, perhaps—many experimenters have opted instead for qualitative research (such as an analysis of readers' own assessments of ease of understanding).

In 1988, Cambridge University psychologist Heather Stark conducted an experiment in which subjects were asked to rate three texts on ease of reading, text coherence and text quality. Reading speed was also measured. The three texts were taken from Bertrand Russell (21 sentences over 4 paragraphs), George Orwell (54 sentences over 7 paragraphs) and Joan Didion (54 sentences over 6 paragraphs). The texts were presented to subjects in one of three formats: with the original paragraphs in place, with no paragraphs at all, and with the paragraph markers moved.

Stark's results will surprise many:

> "... the use or misuse of paragraph boundaries had no measurable effect on subjects' reading rate or ratings of ease of reading, coherence, or goodness ... [It] doesn't seem to make a difference whether a text is explicitly divided into paragraphs or where the paragraph cues occur". (Stark 1988, p. 294)

In other words, the *unparagraphed* versions—even those that were 54 sentences long—were considered just as easy to read, of similar coherence and of similar quality to the paragraphed versions. Stark concludes her paper thus:

> "Given the persistent intuition that paragraph markings make text easier to read, it is surprising that the current study provided no support for this idea. Reading speed and ratings of ease, coherence, and goodness were not affected by the presence of or position of paragraph cues." (Stark 1988, p. 297)

An experiment reported in 1992 found similar results (Markel, Vaccaro & Hewett 1992). Three consecutive paragraphs were taken from a journal article on compact discs, likewise from a journal article on poverty, and likewise from a journal article on fluoridation (giving nine paragraphs in all). For each domain, the text in the three paragraphs was presented to subjects either as (a) the original three paragraphs (each containing about 80 words or 4 sentences), (b) merged into two paragraphs (each containing about 120 words or 6 sentences) or (c) merged into one paragraph (of about 240 words or 12 sentences). Subjects read the blocks of text and were asked to give a Likert response to a number of questions, one of which was "I found it easy to understand the writing". (A Likert response is one from a set ranging from *I strongly agree* to *I strongly disagree*.) The conclusion:

"... paragraph length did not affect the readers' attitudes towards the expertise of the writer, *the ease of comprehension,* or the quality of the passage ... [Paragraph] length is not such a dominant textual feature that it affects ... the ease of comprehension." (Markel, Vaccaro & Hewett 1992, p. 455f. Emphasis added.)

These experiments did not test comprehension, but only readers' assessment of the ease of comprehension. In other words, they sought qualitative rather than quantitative results. Some subjectivity is perhaps unavoidable: some subjects could have been inclined to rate material as easy to understand even though they didn't understand it. Still, the two experiments gave remarkably similar results.

## Another approach

There is a way to measure comprehension of transient reading while minimising the risk that readers will switch to studious reading during testing, namely, by using a cloze test. In a cloze test, subjects are given texts to read in which every fifth or sixth word has been deleted. They are asked to fill in the missing words. The number of correct words entered is a good measure of subject's comprehension of the text.

A cloze test can be used to test whether paragraph length — as measured by the number of sentences — affects comprehension. If Information Mapping — and folk wisdom — is correct, subjects faced with a text composed of a number of small paragraphs should score higher in a cloze test than subjects who are faced with the same information but presented in a single chunk of concatenated text. That is the hypothesis the experiment described below set out to test.

## Materials

A snippet of text was selected that was neither too simple nor too technical. The snippet — taken from *What to do in an emergency,* published by Readers Digest in 1987 — was prepared in two ways:

- extracted as-is: five distinct and consecutive paragraphs, comprising 15 sentences in total (see figure 9.1)

- the same five paragraphs (and 15 sentences) but concatenated into a single block of text without standard paragraph indicators (see figure 9.2).

# What you need in a first aid kit

A home first aid kit _____ mainly intended for minor injuries _____ you can treat yourself, but _____ should also be equipped to _____ with injuries that are more _____ until the victim gets professional _____ help. It should be kept _____ a well-sealed plastic box, _____ as an old ice-cream _____. Put the box on the _____ shelf of the hall cupboard _____ some other place out of _____ reach of children. Do not _____ first aid materials in unsealed _____ in the bathroom or kitchen _____ they may deteriorate in the _____ air. When you go on _____ holidays, take the kit with _____.

Write the address and telephone _____ of your doctor and the _____ of the Accident and Emergency _____ of your local hospital on _____ piece of paper and fix _____ to the inside of the _____ aid box. Tape it to _____ underside of the lid, for _____.

Do not keep old medicines _____ over from a previous illness. _____ them down the toilet or _____ them to the chemist.

First _____ kits can be bought from _____, but you can make up _____ own from the items shown _____ and at the same time _____ familiar with what your kit _____. When buying a first aid _____, check that it conforms to _____ Australian or New Zealand Standard.

_____ bush walks — particularly in remote _____ — take a small first aid _____ which includes a foil blanket (_____ called a space blanket). The _____ can be wrapped around a _____ to preserve warmth in freezing _____. In hot weather it can _____ used with the silver side _____ outwards as protection against the _____ rays.

Figure 9.1  Snippet A, showing paragraph indicators

# What you need in a first aid kit

A home first aid kit _____ mainly intended for minor
injuries _____ you can treat yourself, but _____
should also be equipped to _____ with injuries that are
more _____ until the victim gets professional
_____ help. It should be kept _____ a well-sealed
plastic box, _____ as an old ice-cream _____. Put the
box on the _____ shelf of the hall cupboard
_____ some other place out of _____ reach of
children. Do not _____ first aid materials in unsealed
_____ in the bathroom or kitchen _____ they may
deteriorate in the _____ air. When you go on _____
holidays, take the kit with _____. Write the address and
telephone _____ of your doctor and the _____ of the
Accident and Emergency _____ of your local hospital on
_____ piece of paper and fix _____ to the inside of
the _____ aid box. Tape it to _____ underside of
the lid, for _____. Do not keep old medicines
_____ over from a previous illness. _____ them
down the toilet or _____ them to the chemist. First
_____ kits can be bought from _____, but you can
make up _____ own from the items shown _____ and
at the same time _____ familiar with what your kit
_____. When buying a first aid _____, check that it
conforms to _____ Australian or New Zealand Standard.
_____ bush walks — particularly in remote _____ —
take a small first aid _____ which includes a foil blanket
(_____ called a space blanket). The _____ can be
wrapped around a _____ to preserve warmth in freezing
_____. In hot weather it can _____ used with the
silver side _____ outwards as protection against the
_____ rays.

Figure 9.2: Snippet B, showing no paragraph indicators

Snippet *A* was manipulated so as to make the start of each
new paragraph obvious. Not only was the first line of each
paragraph indented (as in traditional publishing), but
additional space was added between paragraphs (in line with
modern practice).

Every sixth word in both snippets was then removed
(leaving 47 blanks in each snippet). Subjects would be asked to
provide the missing words.

Both snippets were further prepared as a PDF form. This would enable subjects to respond online.

## Method

The following were invited to participate in this experiment:

- students enrolled in *Technical Writing and Editing* at Melbourne University in 2011
- subscribers to *austechwriter* (an internet discussion forum for technical writers)
- participants in technical writing and scientific writing courses held by Abelard Consulting during 2011.

Participants were asked to fill in the missing words (and to leave blank any that were not immediately obvious to them).

When sufficient responses had been received, the number of correct words in each response was calculated. Close synonyms were accepted where the original word was not provided.

The average number of correct answers was then computed for each group of subjects. These averages (or means) were then compared using a $t$-test (for independent samples) in Stata statistical software.[17] All statistical tests were two-sided and a $p$-value $< 0.05$ was considered statistically significant.

## Results

The results are summarised in table 9.1. They show no statistically significant difference between the two means (with $t = 2.54$, degrees of freedom $= 92$ and $p = 0.01$).[18] In other words, the very small difference between the means is just as likely to have occurred by chance as to have been caused by some cognitive mechanism or other.

---

17. *Stata statistical software*, release 11.0, StataCorp, StataCorp, College Station, 2010.
18. My thanks to Dr Gillian Dite, Centre for MEGA Epidemiology, University of Melbourne, for help in analysing the data.

Table 9.1: Comprehension of chunked and unchunked text

|                     | Responses | Mean  | Standard Deviation |
|---------------------|-----------|-------|--------------------|
| A: paragraphs       | 36        | 43.64 | 2.11               |
| B: No paragraphs    | 58        | 44.67 | 1.79               |

This analysis assumes that the samples exhibit normality (that is, they fall within the typical bell-shaped distribution). The data did show some skewing, making it possibly better suited to a non-parametric analysis, such as the Wilcoxon rank sum test. That test gives much the same result: there is no statistically significant difference between the two distributions (with $z = 2.26$ and $p = 0.02$).

## Conclusion

This experiment suggests that there is no loss of comprehension if as many as 15 sentences of material are presented to readers unchunked. This is nearly double the maximum paragraph length recommended by Information Mapping. This is in line with our analysis of transient reading habits described in the previous part of this paper. Readers do not notice the size of the chunk they are reading or, if they do, it does not affect their understanding of it. Chunking, in other words, seems not to be necessary for comprehension.

## So why chunk?

Chunking of informational text into paragraphs is not unimportant. There are many reasons why we do it and should continue doing it. For instance, paragraphing:

- "relieves the forbidding gloom of a solid page of text" (Hudson 1993, p. 294)
- enables writers to present their ideas in logically related chunks

- meets readers' expectations that ideas are being presented in discrete and cohesive chunks
- satisfies the reading practice of the skimming reader (by enabling writers to present each of their main points at an easily identifiable place, namely, the first sentence of each paragraph)
- enables readers to quickly locate text (when paragraphing is combined with titling).

The point of this paper has not been to discount the practice of paragraphing (or at least the practice of creating *well-formed* paragraphs). Rather, it has been to show that *comprehension* seems to be unrelated to paragraph length. We chunk to help readers engage with texts, to help them find the information they are after, to satisfy their expectations about the purpose of paragraphs, and to help them get the essence of what we are saying if they haven't time to read every word. But such chunking doesn't help readers *understand* what they are reading. That is what the research discussed above clearly suggests, namely, that unparagraphed text is considered by readers to be just as easy to read as paragraphed text.

To argue that comprehension is unrelated to paragraph length is not to imply that paragraphs can be of any length. Although the experiment conducted by Markel, Vacarro and Hewett (see page 147) showed that subjects considered unparagraphed chunks of text as easy to comprehend as paragraphed chunks, it also showed that subjects *preferred* that the unparagraphed chunks—those composed of concatenated paragraphs—to have been paragraphed:

> "... regardless of which of the three passages they were reading, [subjects] felt that the 1-paragraph and 2-paragraph versions would benefit from shorter paragraphs, but that the 3-paragraph versions would not." (Markel, Vaccaro & Hewett 1992, p. 455)

This is hardly surprising given readers' expectations about the purpose of paragraphs. A reader who reads a chunk of text formed by the concatenation of three well-formed paragraphs is likely to sense the discontinuity between the merged topics.

They will detect that the chunk is about three distinct topics and wonder why it has not been written according to the well-entrenched convention of one-topic-per-paragraph. So a preference for chunking-by-topic is to be expected. We are habituated to read material chunked in that way.

If writers heed readers' expectation that a paragraph is the container for one topic and that all the sentences in it are related to that topic, then long paragraphs should be rare. Most writers simply don't have more than perhaps 10 things to say about any single indivisible topic. (And if the topic is logically divisible, then it should—if it is to meet readers' expectations—be split into its indivisible parts.) But the important point to note here is that paragraph chunking is better determined by *logic* than *length*. Readers expect *discreteness* and *cohesion* in a paragraph: *one* idea supported by a number of *related* sentences. If a writer happens to have, say, 10 things to say about one discrete idea, the paragraph will need to be ten sentences (or at least 10 clauses) long. To break that one paragraph into two solely on the grounds of the number of sentences in it rather than their logical connectedness is almost certain to distract the reader. The expectation that the second paragraph is presenting a new topic will not be met and the reader will likely be distracted.

To sum up: a chunking limit based on sentence number overlooks the fact that a paragraph is considered by readers to be a *logical* unit. To split a logical unit into two is just as distracting to readers as concatenating different logical units into the one paragraph. Readers' prime expectations about a paragraph—*discreteness* and *cohesion*—are not met. Whatever chunking limit we might apply to a paragraph should not be a fixed number or range (such as Horn's 7 ± 2). It should be a combination of two numbers: one fixed, one moveable. The fixed limit is the number of discrete topics presented in the paragraph. It should be fixed at 1. The moveable number is entirely dependent on what the author is attempting to convey. It is a number that defines the paragraph's degree of cohesion (or how well the sentences are related to a single topic). A sentence count in excess of this limit is what writers should be wary of, not a sentence count in excess of some fixed and seemingly artificial value, such as 7 ± 2. If a paragraph has eight sentences and only six are about the same topic, then the cohesion limit is six and the author has exceeded it by two. That is reason to take a scalpel to the paragraph. If a paragraph has 12 sentences and all 12 are related to the one topic, then the cohesion limit is 12 and the author has not exceeded it. It is fine as it is, and the author should have no fears that it has exceeded some comprehension limit.

In a word: a paragraph should be as short as possible but as long as necessary. It is as short as possible when it is about a single, indivisible topic; it is as long as necessary when everything that the writer wanted to say about that topic is contained within it. To insist on a sentence-number limit fails both logic and science.

## To conclude

Throughout this paper, I have presented proof that:

- Information Mapping—or at least its claim that cognition is limited to 7 ± 2 chunks—is not based on George Miller's research nor on any research that Miller quoted. To claim that it is ignores, or misunderstands, what Miller actually wrote and said.
- Even so, Miller's research has been superseded by more recent studies.
- The way Information Mapping applies the chunking principle is at odds with the way people read texts. The parent or sibling structure of a text is rarely if ever noticed by readers (whether seeking spur-of-the-moment information or in the exceptionally rare case where a reader reads the text all the way through).
- Even if readers did notice the parent or sibling structure of a text, no plausible evidence has been adduced to support the claim that the limited capacity of our *short-term* memory helps us comprehend anything other than

atomic information (such as a clause or sentence). At the level of chapters, sections, maps and blocks, the capacity of our short-term memory appears entirely irrelevant.

- To understand a chunk is to understand the sentences in it and such understanding is unaffected by the number of sibling sentences that have to be read.
- Research in cognitive psychology—by Kintsch, Stark and others—shows that the number of sentences in a paragraph does not affect a reader's comprehension of that paragraph, contrary to the claims of Information Mapping and the recommendations found in many language handbooks.
- Chunking is valuable but is better based on chunk discreteness and cohesion than on chunk size. Logic, not length, is a better guide to quality paragraphing.

# 10: A lament for the vanishing index

What a whirlwind of technological evolution has swept through the last 40 or so years: desktop computers, mobile telephony, cloud computing and, of course, the web. Not to marvel at The Age of Computing and its manifold spin-offs is, surely, a sign that one is lacking the genes that stamp us as *Homo sapiens*.

But silver linings sometimes harbour clouds. Not everything born of The Age of Computing deserves unqualified admiration. The democratisation of publishing might well be extraordinarily empowering, and continue to change the way we live for the better. But it is not difficult to see some downsides, such as cyber-addiction, the ease with which the meek can be bullied online, and the whittling away of something we will later regret losing, namely privacy. As Shakespeare wrote, no doubt in a melancholy moment, "all that glisters is not gold".

Of more relevance to the work of technical writers, the democratisation of publishing—and in particular the ease with which everyone's musings can be disseminated—adds a further layer of complexity to our craft. Language use has never been fixed, but the rate at which it is changing has reached a new high. Anomalous usage, once largely stillborn, can now quickly become a fad, a fad a trend and a trend a new convention. (Look, for example, at the birth, and quick maturing, of the *open* hyphen.) Only those deaf to language

First published in *Newsletter*, the bimonthly publication of the New South Wales branch of the Australian Society for Technical Communications, July 2012.

will fail to recognise that English has changed over the last 20 or so years.

Why is that a problem for technical writers? Well, are we not obliged to communicate in ways that require the least effort on the part of our readers and cause them the least distraction? That lies at the heart of the philosophy behind our oft-repeated mantra that we should write in an "audience-centric" way, that we should be the "readers' advocate". Thus we need to write in ways that are maximally *familiar* to our audience, and this implies that we keep abreast of changes in language. (It would distract many a contemporary reader if we stuck to the so-called laws of language etched in our minds by the gerund-grinders of yesteryear.) In stirring up mud — among the occasional gold flecks — in the stream of general communication, language-change imposes an additional, a weightier, burden on writers who need to deliver transparent, immediately digestible meaning. Thus technological change is not necessarily an *unqualified* blessing. Good though it might be, no end of the beam balance is resting on the bench, proud and satisfied.

Like most, but perhaps more so, technical writers are fascinated — and some possibly bewitched — by new technology. It bursts through the dull grey, the stifling monotony, of quotidian routine, and ignites our narcoleptic imaginations. And when new technology comes with a promise of money to be saved, it is hard not to be blinded to the downsides. Some changes following in the wake of new technology are impossible to resist. Language change born of an anarchic blogosphere falls into this category. But other changes can be resisted: if not the technology itself, then at least its application to areas of dubious benefit.

Take, for example, online learning, a relatively new technology that promises to open up knowledge to all. Think of knowledge without borders and the consequent prospect of a New Enlightenment. But then think too of the narrow, pre-defined avenues of exploration it offers, and the vanishing of extemporaneous learning and instantaneous clarification.

Only misguided educationalists (and miserly governments) could consider the one-person classroom—aka the screen—to be a superior learning environment. It is at best an adjunct, not a substitute for teacher-led learning.[1]

Think of structured authoring and the promised simplicity of format-free writing: but then think too of how it shackles instructional creativity and denies the writer a chance to exploit the communicative power of textual appearance.

Think of the customer-created wiki with its democratic invitation to *Everyperson* to share their wisdom: but then think too of the distracting anarchy of styles, the unexamined guesses camouflaged as truth, the unthought-of gaps, the shaky reliance on the generosity of others, none of which is found in texts meticulously sculpted by technical writers.

Think of content management systems, with their tidy, *waste-not* philosophy that enables every snippet of text, every table and figure, to be re-used and repurposed: and think too of the clashing styles, the inconsistent vocabulary and the clanking transitions that bedevil many mash-ups sewn together from such op-shop leftovers.

Finally, think of the full-text electronic search and the speed with which it delivers information to the reader: and then think too how the ease and speed of its implementation has led to the near-demise of a once essential adjunct to self-paced learning—the humble index—without which the branch we occupy in the gnarled tree of cultural evolution would be more brittle and no doubt stunted. (Can anyone honestly attribute their educational achievements to reading texts that lacked an index? A corollary: is our current default position of omitting

---

1.  A study reported in *Scientific American* found that "reading online may not be as rewarding—or effective—as the printed word. The reasons: The process involves so much physical manipulation of the computer that it interferes with our ability to focus on and appreciate what we're reading; online text moves up and down the screen and lacks physical dimension, robbing us of a feeling of completeness … The visual happenings on the screen and your physical interaction with the device is distracting …" See "Online v. print reading: which one makes us smarter?", *Scientific American*, 23 December 2008.

indexes condemning future readers to an education far less rich than we enjoyed?)

So, what has happened to the index? Let's look at some claims made against indexes: they are too expensive; they add to the production time and thus can delay a product's release; and a full-text search facility is far better than an index.

Is an index expensive? Imagine a 200-page user guide. By Australian standard AS4258, such a manual would take a technical writer about 600 hours to write. At current contract rates, that would cost about $40,000. Now a 200-page manual would take, at most, 30 hours to index. This is one-twentieth as long as it took to write the manual. So indexing would add about $2,000 to the cost. This is chickenfeed to most companies, and an amount more than offset by the goodwill an index will generate in readers of the manual.

But need such an index take 30 hours to compile? Well it might if the manual was sent to an indexer who knew little beforehand of its contents. But there is an alternative: get the technical writer to index—*and index as they write*. Rather than indexing only when the writing is finished—an oddly common practice—indexing *while writing* would see the cost of indexing largely absorbed in the cost of drafting (for it takes very little time for a writer to insert an index marker as they go). Moreover, the resulting index is likely to be superior, and for two reasons: (a) during the struggle of drafting, the writer will no doubt be assessing various synonyms and thus, given their intimacy with the text, will have a better idea than an external reader (such as an indexer) of what words are worthy of indexing and (b) the index can then be added to the purview of whoever reviews the manual during its development (thus leading to incremental improvements in the index).

Another objection is that an index adds to the production time and thus can delay a product's release. That might be so if indexing is done post-drafting. But if the technical writer indexes as they go, little time is added to the schedule. Even so, why can't documentation development begin earlier if that would be needed to accommodate indexing?

But the most common objection to including an index in technical documents is that such documents are mostly provided in electronic form, and the most common forms — PDF and web help systems — offer a full-text search facility, a facility supposedly superior to an index. But is it superior? The standard by which this needs to be judged must be that of *usability*, and one of the three pillars of usability that underpins all informational writing is that information must be *easy to find*. So, is information found via full-text searching facility found more *easily* than if it were found via a human-created index? The answer is surely no, and for a number of reasons.

Firstly, a search facility is of no use to the many who — in recognition of the fact that reading on screen is slower, is subject to the siren-call of distraction and leads to lower comprehension — print out documents.[2]

Secondly, a search facility could leave a searcher thinking that a particular topic is not covered in the text — even though it is — if the searcher happened to be searching on a *synonym* of the term that the author actually used. For example, searching for *Basedow's disease* in a medical manual written outside Europe is likely to yield no hits, as Basedow's disease is known as *Grave's disease* elsewhere. Same disease; different names. Or searching for *kerosene* will yield no hits if the author's preferred term is the British equivalent *paraffin*. A human-created index will alert the reader to synonyms with a *See* cross-reference: *kerosene* See *paraffin*. (Likewise with spelling variations: searching a document for *tyres* might yield no results if the author preferred American spellings.)

Thirdly, a human generated index will alert the reader to *related* topics by way of a *See also* cross-reference: *epilepsy* See also *seizures*. The reader, unsure about the terminology used in the document, might well find what they are looking for only by going to a cross-referenced topic. Such cross-referencing is not found in full-text searches.

---

2. Ackerman and Goldsmith (2008) reported that "Survey respondents at all ages (17–61) reported that they usually print digitalized texts before studying them".

Fourthly, a human-generated index can suggest to the reader where most information about a topic can be found. This could be indicated by the presence of *span* of page numbers (45–51) as opposed to a single page number, or because the page number is set in a different style (such as bold). At best, a search engine might give a hit a ranking, but this is often a blunt indicator (being little more than a measure of how often the search term appears at a particular location).

Fifthly, an index is a better guide than a table of contents to the sweep and depth of knowledge available in the text, revealing far more opportunities for discovery and learning.

Sixthly, I am going to be stuck if I am not sure of the spelling of the term I am looking for. For example, if I enter *antidiluvian* as my search term rather than *antediluvian*, I will, in all likelihood, get no hits. A quick glance at an index would alert me to the correct spelling — and to where the term appears in the document. Some search engines offer *fuzzy* searches and these might overcome this particular problem for some terms. For example, the fuzzy search available with the online version of the Macquarie Dictionary displays *antediluvian* if I enter *antidiluvian*, but it gives no suggestions when I enter *Xrays* or *anoreksic*, and it gives the wrong suggestions if I enter *mison* when looking for *meson*.

Finally, a full-text search can lead to numerous time-wasting partial hits. Having found no hits using, say, *income reports* as my search term, I might then search on *reports*. But before I get a hit on *revenue reports* — the author's preferred term, unbeknown to me — I might have to skip over scores of irrelevant hits (*sales reports, shareholder reports, debtors reports* and so on) before, a hundred or so pages later, I get a hit that is relevant. A human-generated index would have entries for *reports: revenue* and *income reports* See *revenue reports*, enabling me to quickly see the author's preferred term and go straight to the relevant topic.

In all these cases, the typical back-of-book index (even one without page numbers, as in an online help system with hypertext entries) provides greater usability than a full-text search. Information will often be found more quickly, where it is found at all. Thus a claim that a full-text search facility is far better than an index needs a measure other than usability to justify it. But what might that be?

It is sometimes said that our clients always want to maximise efficiency and that shedding indexes aids us in achieving that goal. Now efficiency might be a laudable goal, but if it were our sole goal—if our readers' needs were not part of the equation—then it could be met quite easily by writing pure, unadorned text: exactly what you get from Notepad. Most clients would baulk at that. They know that readers of product manuals would naturally, even if unfairly, generalise their first impressions: shoddy manual, probably shoddy product. At the other end of the spectrum is the incorporation of every bell and whistle ever imagined in tech-writer heaven. Most clients would baulk at that too. The cost of producing such documentation would no doubt outweigh the accumulated benefits. So somewhere along the spectrum from maximising efficiency to delivering the greatest usability to the greatest number is a client's sweet point. This point won't be the same for every client, and the problem is that most clients don't know where their sweet point is.

And in this ignorance lies an opportunity that is often wasted: the opportunity to educate a client about what end-users want (and thus an opportunity to give end-users what they need). Instead, some of us gratefully, sheepishly, do just what a client asks, even if we suspect that the client has thought little, or knows nothing, about usability in documentation. We forego the opportunity to up-sell, to truly be the readers' advocate that we endlessly brag about, perhaps thinking that a suggestion to include, say, an index might be interpreted as nothing more than a wish to fatten our own purse. This is an area where we need to lift our game, to fine-tune our rhetoric, so as to better convince our clients that giving end-users what they need is more than just a win for the technical writer. It is also a win–win for client and end-user. A win–win–win all round, in fact.

# 11: Writing for an international audience

## The INTECOM guidelines

In 2003, the International Council for Technical Communication (INTECOM) published guidelines to help technical writers write English-language documentation for an international audience (see INTECOM 2003). The guidelines make two main recommendations:

1. US spelling and usage are recommended for documentation that will be read primarily in countries where US spelling and usage are prevalent, and British spelling and usage are recommended for documentation that will be read primarily in countries where British spelling and usage are prevalent or were part of the country's history.
2. US spelling and usage should be used in English-language documentation that will have worldwide use.

Before considering these recommendations, let's formalise a fundamental principle of communication, technical or otherwise, a principle I'll call the *principle of maximal familiarity*. This is the principle that, to minimise distraction in communication, we should attempt to communicate in the language that is maximally familiar to our intended audience: familiar spellings, familiar word choice, familiar idiom, and so on. This principle underpins all good technical writing and is implied by the audience-centric approach that is the cornerstone of our profession. So how do the INTECOM recommendations fare in the light of the principle of maximal familiarity?

First published in *Tech Talk*, March 2004.

## Recommendation 1: Adopt the prevalent English

On the face of it, recommendation 1 might pass a rushed muster. If our intended audience is, say, British automotive engineers, we would, as technical writers, adopt British English in our user documentation. And the guidelines provide a useful glossary of distinctly British terms to help non-British authors, contrasting these terms with their US equivalents, as in "car (Br); automobile (US)".

But to the authors of the guidelines, recommendation 1 means much more than just writing in British English for a British audience, and American English for an American audience. They divide countries where there are English speakers into two camps:

1. Countries where US spellings and usage are prevalent, being North and South America, the Philippines, Japan, China, North Korea and South Korea. (Let's call English speakers in these countries A-people.)
2. Countries where British spellings and usage are prevalent or were part of the country's history, being the UK, Australia, New Zealand, South Africa, many Caribbean countries, India and Pakistan amongst a few others. (Let's call English speakers in these countries B-people.)

And now comes a curious recommendation: if writing for A-people, we should use US spelling and usage, and if writing for B-people, we should use British spelling and usage:

> "If writing for an audience solely in the UK, the Scandinavian countries, Australia, New Zealand and South Africa, then British usage is appropriate."

This is extraordinary advice. Since Australia is a B-people country, the advice, if followed, would have a technical writer adopting British spellings and usage if writing for an Australian audience. This means that the writer would write *current account* rather than *cheque account*, *pavement* rather than *footpath*, *motorway* rather than *freeway*, *dustbin* not *rubbish bin*, *antagonize* not *antagonise*, and so on.[1]

This advice is at odds with the principle of maximal familiarity. British spellings and usage might be *prevalent* in Australia (to use the language of the recommendation) but only in the sense that there are more similarities than dissimilarities. But as numerous published guides to Australian English attest, there is a distinctly Australian English, and its distinctiveness is sufficient to warrant its adoption when we are writing for an Australian audience.

Similarly, there is a distinctly Indian English, a distinctly South African English, a distinctly New Zealand English, and so on. A recommendation to adopt only British English despite this heterogeneity wrongly belittles the legitimacy of the numerous other variants of the English language.

## Recommendation 2: Adopt US English for international documentation

The advice to those whose work might be read worldwide is equally unconvincing:

> "the Project Group [that is, those contributing to the guidelines] has mostly suggested using US spelling and usage for English-language documentation that will have worldwide use. Our rationale is simply that people who are accustomed to US spelling practices find British spelling to be strange or quaint, or may even think the writer cannot spell correctly. On the other hand, most people who use British spelling and usage have also been exposed to US spelling and usage, so that even though they don't use it themselves ... they recognize it and more readily adapt to it." (INTECOM 2003, p. 3–4)[2]

Just to underscore how extraordinary this advice is, note that no part of its so-called derivation is based on how many A-people and B-people there are. For this reason, the advice is, in effect, tantamount to saying that, regardless of whether A-

1. The British words, and their spelling, are taken from the glossary provided in the guidelines.
2. No reference to any empirical, statistically-significant, research is given to support this claim.

people are in the minority, the linguistic practices of A-people should prevail over those of B-people when you are writing for an international audience. And this is simply because A-people are less tolerant of linguistic diversity than B-people. This is starkly at odds with the principle of maximal familiarity. It is also crudely dismissive of A-people: are they really that ignorant, that intolerant of diversity, that they must be kept isolated from foreign tongues, even English ones?

## Another approach

INTECOM's recommendation 2 addresses an uncommon scenario. Few technical writers ever need to write for a worldwide audience. A much more likely scenario is writing for an audience composed of speakers of a limited number of English variants. An example is writing a user guide to accompany a product that will be sold in Australia and in the USA. How should we decide on the English to use if localisation is not an option?

One argument might be that since there are more US-English speakers than Australian-English speakers, we should write the manual in US English. This might sound fair, but would such an approach always accord with the principle of maximal familiarity?

Suppose that the company in question expects to sell 10 000 units in Australia but only 1000 units in the USA. If we based our decision on what English to use solely on population size, then 10 000 customers would encounter an English that is not their own, and only 1000 an English that is. But if we apply the principle of maximal familiarity — with the democratic imperative that it implies — we would choose to write the manual in Australian English, since 10 000 would then encounter language that is their own and only 1000 would not.[3]

---

3.    And to fully maximise familiarity, we might, in addition, choose to use international words for words that might be understood only in the majority language (for example, *fuel* rather than *petrol*).

This method has general application: every exporting business has, through market analysis, a good idea of how many units of a product they expect to sell in other countries. (No business would manufacture products for export without a clue as to which countries they were to be exported to.[4]) So, the technical writing team needs only ask the marketing team which countries the product to be documented is be exported to and how many units they expect to sell in each such country. An application of the principle of maximal familiarity, in the manner outlined above, would then determine the variant of English to use (which is not to say that the task will always be as simple as in our example).[5]

The principle is no less relevant to worldwide documentation, that is, to documentation that the authors expect will be read in every country where there are English speakers. The authors could ascertain the number of English speakers in each country and the characteristics of the particular variant of English used before deciding on the variant or hybrid of English that would give maximal familiarity. Had the INTECOM team concentrated on gathering such information—instead of relying on spurious national stereotypes—the results of their efforts might have been of wide-ranging value: not just to writers, but to all who have an interest in the many-branched evolution of the English language.

## Conclusion

It is an axiom of technical writing that we write for our audience: we give them what they need to know, based on

---

4.    Even companies that sell via the web will have a fair idea of, or can make educated guesses about, the sales they are likely to make in countries with English speakers.
5.    Indeed, in some cases, the English rightly chosen will not be the English spoken in the highest-selling country, for expected sales in the highest-selling country might be out-numbered by the expected *total* sales in a number of lesser-selling countries whose *shared* language is different to that used in the highest-selling country.

what we know they know, and in a language that informs with minimal distraction. Language cannot inform with minimal distraction if it employs words, spelling, punctuation and idioms that the intended audience is not accustomed to. And thus is derived the principle of maximal familiarity.

By giving insufficient weight to the fact that there is a multitude of legitimate English variants, the advice that we should shoehorn our language into just one of two variants is obviously at odds with the principle of maximal familiarity. So too is advice that we should give primacy to the customs of one language group over all others irrespective of the size of that group in our total audience. Hence the INTECOM guidelines are best ignored if we are to maintain our customary audience-centric focus. Instead, we should attempt, wherever practicable, to quantify our likely audiences and their particular language customs, and then submit our findings to the calculus of maximal familiarity.

## Addendum: Diversity in contemporary Englsih

A fair assumption might be that the study group behind the NTECOM guidelines did not have an appreciation of the diversity of contemporary English. So let's just dip into the sea of Babel that is contemporary English.[6]

Consider firstly vocabulary and meaning. In American English a *bill* is what a speaker of Australian English would call a *banknote*. Similarly, an American *drugstore* is a *chemist* in Australia. A *period* in American English is a *full stop* in most other Englishes. A *lay-by* in Australian English is a purchasing arrangement while in British English it is a part of a road where vehicles can pull up out of the stream of traffic. Likewise, a *capsicum*, *footpath* and *nature strip* in Australian English are called *pepper*, *pavement* and *verge* in British English.

There are also intra-country variations. For example, in most states of Australia, the pole that carries electricity and

---

6. For a sketch of some of the many variances in English throughout its long history, see Marnell 2015, pp. 58–63.

telephone cables is called a *power pole*; but in South Australia it is more commonly known as a *stobie pole*. Another regional variation is the name for German sausage which, depending on where you live, might be called *fritz*, *devon*, *polony* or *strasbourg*. *Flake* in Victoria is called *shark* in Western Australia, and a *potato scallop* in New South Wales is called a *potato cake* in Victoria. Finally, a *suitcase* is still called a *port* by many residents of Queensland.

What is called a *beach house* in Australia is called a *crib* in the South Island of New Zealand, but a *bach* in the North Island. A *sandwich* in the south of England is a *butty* in certain northern counties; and a *turnip* south of Hadrian's Wall is a *neep* north of the Wall. Further, gym shoes (or trainers) are variously called throughout the British Isles *plimsolls*, *sandshoes*, *pumps*, *gollies*, *daps*, *whiteslippers* and *gutties* (Trudgill 1999, p. 110).

There are also grammatical differences. In the south-west of England it is common to say *I did go there every day* whereas most other speakers of British English would say *I went there every day* (Finegan et al. 1992, p. 359). Double negatives are not a feature of the language of south-east England (where, for example, the preference is for *I don't want any trouble*) whereas the double-negative is preferred in most other parts of England: *I don't want no trouble* (Finegan et al. 1992, p. 87).

Grammatical differences can also be seen in other variants of English. For example, in most of Australia the past tense of *to bring* is *brought*, but many in Far North Queensland prefer *brang*. As for American English:

> "Instead of *I saw it*, a New Englander might say *I see it*, a Pennsylvanian *I seen it* and a Virginian either *I seen it* or *I seed it* ..." (Quirk et al. 1972, p. 14)

There are numerous spelling differences between American English and other Englishes. Americans typically drop the *u* from words ending -*our*, so that *colour* in British English is *color* in American English. There is less letter-doubling in American English: *labeled* is preferred to *labelled*, as is *canceled* to *cancelled* and *modeled* to *modelled*. (There are some words, however, where letter-doubling occurs only in American English: *enroll* and *fulfill,* for example.) American English prefers *-ize* to *-ise* endings in such words as *organize, colonize, authorize* and so on. This was the preference in Elizabethan England, but it is not so today. In Australian and New Zealand English, these words have always had an *-ise* ending.

Finally, punctuation. Inter-country variation is less common than it is with pronunciation, vocabulary and spelling. The most notable variation is the widespread use of the serial comma in American English and its near absence in most other Englishes. A serial comma is a comma before the final *and* in a run-on list (that is, in a series of listed items). The second comma in the following sentence is a serial comma:

> The flag is red, white, and blue.

In British and Australian English, the serial comma is rarely used. It is added to a sentence only if ambiguity would result without it, as in the following example:

> This page of the intranet covers departmental policy, organisational charts and resource plans and purchasing templates.

A serial comma would make clear what the second and third items covered on that page of the internet are. At present it is impossible to tell.

It should be clear that the differences between the various Englishes—together the sovereign right of every English-speaking country to use its English in whatever way it pleases—strips all validity from attempts to foist one of just two Englishes on every English-speaking country.

# 12: Controlling technical vocabulary

Clarity is undoubtedly one of the many hallmarks of effective technical communication, and one feature that contributes to clarity is consistency in the use of words or terms. This can be difficult to achieve, especially in long or multi-authored works. A decision to use a particular term to describe a particular thing must be applied without exception, and this can require a good memory and maybe successful collaboration.

## The principle of single and distinct denotation

That there is good reason for adopting consistency in word choice—and therefore avoiding what is known as *elegant variation* (Fowler 1965, pp. 148–151)—is easy to demonstrate. If a reader is presented with *two* terms for *one* unfamiliar thing, the impression could be created that two distinct things are being discussed, not one thing in two ways. And this is an obvious impediment to swift and direct learning. For instance, if you are learning a word processing application with the help of a user guide and find, after a description of the tool bar, a direction to click on the tool bar rendered as "click on the tool ribbon", you may well wonder where the tool ribbon is despite having just been shown where the tool bar is. The context might help, but then again it might not (having been left unnurtured by the overriding demands of deadline or budget).

First published in *Keyword: A journal for technical and scientific communicators*, vol. 8, no. 1, 1998, pp. 14–17.1

This type of potential for confusion is not limited to specialist vocabularies. For example, writing "the code is *allocated* by the system" and then later, when referring to the same thing, "after the code is *assigned* ... " is just as likely to cause readers to wonder whether two things are being referred to. (Perhaps the codes that the system *allocates* need to be *assigned* beforehand — in hard code or in a lookup table — one to each of the possible things that may need to be coded by the system.)

Even where there is no confusion, or confusion is transient, synonym-twinning — that is, using more than one term to denote the one thing — creates an unnecessary burden on the learner, namely, the need to master a vocabulary that is larger than necessary. Such a burden renders the documentation less than maximally effective.

Despite these considerations, there are instances where synonym-twinning can be useful. For a start, there are many strong synonyms in such common use that their intermixing would cause the reader no problem. Indeed, their intermixing would add variety and vigour to the writing, helping to keep readers engaged with the text. Synonym pairs such as *but* and *however*, *hence* and *therefore* and *assume* and *suppose* fall into this category. But writers still need to be cautious lest, unknown to them, two common strong synonyms are, in a particular context, read by some readers as referring to different things. Consider this step in a procedure:

> From the **Transform** menu choose **Matrices** and then select **Trace** from the sub-menu.

Many readers would take *choose* and *select* to be well-entrenched strong synonyms, yet it is conceivable that some readers might initially think, if only briefly, that one directed a left mouse click and the other a right.

Another situation that might benefit from synonym-twinning is where a new term for a common concept is considered necessary and, to enable familiarity to aid learning, the common term is used in addition to the new term. An example is "*category axis* — also known as the *x-axis* — ...". Nonetheless, the familiar term paired with the preferred but possibly novel term is usually avoided in the remainder of the documentation.

The case is similar with the contraction to, or expansion of, shortened forms (such as abbreviations, acronyms and other types of diminutive). Once the contraction or expansion has been made, the writer ideally sticks to one form (usually the contracted form, but sometimes the expanded). Thus we read:

"The cause of Creutzfeld—Jakob disease (CJD) is unknown ... In Britain, CJD has been found in ...".

With these few exceptions—enlivening with harmless non-technical variety, briefly pairing a new term with a familiar term, and providing or expanding a shortened term—good technical communication favours single denotation, namely, the consistent use of one term in referring to one thing.

The companion principle to the principle of *single* denotation is the principle of *distinct* denotation. This latter principle prescribes the use of different terms for describing or referring to different things. In this case, the writer is not so much trying to prevent readers conceiving of two things when only one is meant, as trying to prevent readers from thinking that distinct things are one and the same. In a word: they are trying to prevent *lexical ambiguity*.

The following example of non-distinct denotation has been abstracted from actual documentation describing a suite of systems one of which generates a software record in response to a customer using a service (a record that is eventually translated into a charge on the customer's bill). A companion system traps records that may have been generated in error, generated incompletely, or with abnormal values. Staff have to investigate each trapped record and delete, correct or accept it. A record corrected or accepted is then "released" back to the charging process, whereas a record found to have been generated in error is "deleted".

All this seemed clear enough in the documentation—until the author began discussing the turnover of records passed to the record-trapping system. "Turnover" referred to the number of records "released" from the system each month, but "released" now meant "accepted, corrected or deleted". In other words, the author was using the term "released" to mean both (a) the return of a trapped record — possibly modified, possibly not — to the charging process and (b) the extraction, purging, removal or whatever of a record from the record-trapping system (which encompassed deleting records as well as returning records to the charging process). Indubitably, some, maybe many, readers will have been confused by this case of unnecessary non-distinct denotation: one word, two things.

Occasionally, non-distinct denotation is unavoidable. For instance, in a poorly designed computer system, one sometimes finds distinct fields with identical literals. (There might be, say, a **Type** field for the type of customer and a **Type** field — maybe even on the same screen — for the type of product purchased by the customer.) In describing such a system, documentors will be compelled, by their obligation to describe precisely what users will see, to use one term for two distinct notions. (In such cases, procedural steps will need to be fattened with geographical instructions.)

Putting aside these few exceptions and unavoidable limitations, maximally effective technical communication requires adherence to what might be called the *principle of single and distinct denotation* (hereafter PSDD). Single denotation facilitates the associating of description with what is described, while distinct denotation minimises ambiguity. But how, in practice, can technical writers adhere to the PSDD?

## Subject-specific thesauri

One solution is to create a *subject-specific thesaurus* (or use one, if one covering the system, product or subject you are writing about has already been created). In its simplest form, a subject-specific thesaurus is list of word pairs, with one word from the subject in question (avionics, electrical engineering, economics or whatever) and its pair being the word the writer or writing team has chosen to use.

Subject-specific thesauri are a little different to the more common generalist thesauri, such as Roget's justly famous thesaurus. Whereas Roget set out to include every current English term, a subject-specific thesaurus sets out just the vocabulary that is common to a particular subject, product, system or industry. Furthermore, while a traditional thesaurus of the Roget variety refrains from giving direction to the reader to adopt any particular term over another, a subject-specific thesaurus usually specifies, for each set of like terms, a preferred term. Ideally, a preferred term is the term that, for a specified audience, most clearly expresses the meaning of the concept represented by the set of like-meaning terms.

As in traditional thesauri, each entry in a product- or subject-specific thesaurus is introduced by a keyterm. The preferred term for a keyterm that is not itself the preferred term is usually introduced by the label "USE". For example, an entry in a subject-specific thesaurus that reads:

**coil clip** USE spring clip

indicates that *spring clip* is the preferred term for *coil clip*. If a keyterm is not followed by "USE" and some alternative term, then the keyterm itself is also the preferred term for whatever concept is represented by it. An excerpt from a product-specific thesaurus is shown on page 186.

A list of keyterms and preferred terms is a minimalist subject-specific thesaurus. Such a list could help writers apply the PSDD and thereby control their language for the sake of greater communicative efficacy. But subject-specific thesauri usually provide considerably more information than this. For instance, they will often include what are known as *scope notes*. Scope notes give the reader an idea of the particular sense of a given keyterm (and can, if given enough attention, enable the thesaurus to double as a glossary or subject dictionary). They are usually labelled SN in a thesaurus, as in:

**release (vb)** USE action. SN: If referring to the removal of records from the error-trapping system.

Fully-fledged subject-specific thesauri — of the type favoured by records management people and library cataloguers — also provide, for each keyterm, words for concepts that are logically related to the concept expressed by the keyterm. Four types of logical relationship are commonly recorded: synonymity, greater generality, greater specificity and general relatedness. The terms instantiating these relationships are known as *synonyms, broader terms, narrower terms* and *related terms* respectively.

A synonym (usually indicated by SY) is a term having the same meaning as, or being very close in meaning to, the preferred term. A broader term (BT) is a term representing a concept of which an instance of the preferred term is a type or sort. A narrower term (NT) is a term representing an instance or example of the concept represented by the preferred term. Finally, a related term (RT) represents some instance of the broader term other than the preferred term.

An example of a complete entry in an architectural thesaurus might be:

> **abode** USE residence. SN: When referring to a place where a person might live for some time. SY: home. BT: building. NT: house, flat, unit. RT: factory, church, hall, shop, gallery, office, hotel, motel.

Such an entry indicates that *residence* is to be preferred to *abode* providing that the sense is a place where someone might live for some time. A common synonym is *home* (although *residence* is still the preferred term). A broader term is *building* (since a residence is a type of building) and narrower terms are *house, flat* and *unit* (since these are all types of residences). Finally, the related terms refer to buildings that are not residences.

By providing writers with instances and examples, a fully-fledged thesaurus of this sort is a useful aid in setting contexts and associating the unfamiliar with the familiar.

Subject-specific thesauri are also useful for controlling troublesome terms that are not especially technical (or which do not have anything to do with the subject of the thesaurus). For example, certain words prefixed with *bi-* are troublesome words and might warrant inclusion in a subject-specific thesaurus. (Will your audience understand *bimonthly* to mean twice a month or every two months? Similarly, an American audience might benefit if material written for it had been controlled by a thesaurus that decreed that *every two weeks* was the preferred term to *fortnightly*, since *fortnightly* is not common in American English).

To sum up: a subject-specific thesaurus helps writers to apply the principle of single denotation by directing them to use one term (the preferred term) where there are a number of terms to describe what needs to be described. A well-constructed thesaurus will also acknowledge the principle of distinct denotation by providing, where practical, preferred terms that denote one thing and one thing alone (or, where this is not practical, providing—via scope notes—a prompt for writers to consider the necessity of spelling out the context of a potentially ambiguous term).

## Opportunities to control language

Documentors are rarely invited to participate in the design of new products and systems. Mostly, they join a project when what they are to document has pretty much taken shape. By this time there is already current a sub-language peculiar to the thing to be documented. One might wonder, then, what influence a documentor joining the project at such a late stage can possibly have on the language already in place. Such doubt overlooks three important facts:

1. The sub-language of designers, analysts, engineers and the like rarely encompasses everything that end-users need to know. There will be some non-technical material that documentors will need to write—such as business context, policies, procedures and troubleshooting hints—which might be expressible in that sub-language.

2. Where the language directed at users needs to be technical, the language adopted does not always need to be—and in many cases shouldn't be—the language of the

designers, analysts and engineers (namely, the sort of technobabble one commonly finds in design documents and functional specifications). There may, for instance, be unlabelled fields on screens that have punishing names in the accompanying functional specification, names which experienced documentors will be quick to ignore in preference to plain English alternatives.

3. The technical vocabulary that documentors encounter is often connotationally distracting, ambiguous or vague, having arisen without concern for the needs of everyday, common-or-garden users (without a concern, in other words, for the PSDD). If this is acknowledged within the organisation, there may be a willingness to accept an alternative, less troublesome vocabulary.

There is, then, room for documentors to choose language even where a technical vocabulary already exists. If documentors have a choice, then it is important that the choice be based on solid principles of technical communication, one of which is the PSDD.

## Compiling a thesaurus

There is no set way to compile a subject-specific thesaurus, although top-down (or decreed) thesauri are less likely to succeed in comprehensively covering the language of any but the smallest projects. More often it is only when writers begin their work that the extent of the language that needs to be controlled becomes apparent. Hence decisions on what should be the preferred term often need to be made on the fly and from the bottom up. This doesn't mean that a project won't benefit from having a thesaurus of core terms in place before writers begin the bulk of their work. But it does mean that processes need to be set up to enable writers to record as they go terms — technical and otherwise — that need controlling.

The terms proposed for controlling need to be reviewed regularly (and, during the early stages of the project, frequently). The quality manager, documentation designer or documentation team leader might be charged with reviewing these terms and deciding which are to be preferred and how each term relates to other terms. This process may benefit from being done in concert with the client, who might have a special interest in pushing a particular language or style of language. It will also benefit from the input of the writers on the team, for their initial unfamiliarity, and their special training, makes them best placed to draw out the weaknesses of a sub-language, weaknesses that those familiar with its vocabulary, special contexts and arcane connotations often fail to see.

## What makes a term a preferred term?

What makes a term a preferred term varies according to whether there are already terms available that describe what needs to be described or whether you need to invent a term. In most cases the preferred term from a set of current terms will be the term with the widest currency (for the term with the widest currency will, usually, be the one understood by most people). Four qualifications to this general rule are necessary if our goal is to maximise the effectiveness of our documentation and meet our client's expectations:

1. Jargon that is likely to mislead or obscure the meaning to novices may need to be rephrased in plain English even though it has wide currency amongst the likely initial audience for the documentation. It may not always be possible to provide a suitable succinct replacement term; but where one can be found, this should become the preferred term. An exception might be where the jargon is so well-entrenched that it will be difficult to usurp.

2. Where one term is widely used to refer to two distinct things, it may be wise to use another term for one of those two things (providing, of course, that the new term is more immediately indicative of whatever it is meant to denote). In such a case, the new term may become the preferred term despite its relative scarcity or newness (although writers may need to include an initial cross-reference to the term that is most widely

used at the moment). The important consideration in deciding to prefer a new term is how likely it is that the intended audience will be misled or baffled by the potential for ambiguity implicit in the various uses of the one term. (This qualification is simply a prescription to adopt the principle of distinct denotation.)

3. Where a common term is used with grammatical or semantic ineptness and that use is likely to lead to confusion, a thesaurus may suggest a less-common replacement term. (Again, how well-entrenched the term is may limit one's success in promoting a new term. For example, preferring *local network* to the redundancy-exhibiting *local area network* is unlikely to succeed.)

4. A particular term may have wide currency—and be particularly fitting—and yet the client may demand another term. (This might be at the request of the client's marketing department, for whom connotation as much as denotation and currency is an important factor.)

The decision to make a term a preferred term may, then, be descriptive (What term has the greatest currency?) or prescriptive (What term exhibits the greatest clarity and is least likely to lead to impeded communication?). Prescriptive judgements must, of course, be made with care, as some term may be so well-entrenched in a particular sub-language that a recommendation to use some other term—one of greater discriminatory power or linguistic respectability—may ultimately prove fruitless and cause, along the way, the very confusion it was designed to avoid.

Where a neologism is needed, considerations such as the following are useful in choosing a preferred term:

- absence of distracting connotation
- metaphorical aptness
- consistency with the client's wishes
- distinct denotation (within the subject area)
- clear indication of the thing the term is to denote
- appropriateness to the language skills and knowledge of the intended audience.

# Conclusion

Whatever the product, the benefits accruing from quality-inspired extras are not always immediately apparent. This is equally true of the benefits of controlling the language used in technical documentation. They are there nonetheless, benefits such as:

- less call on help desks and customer service staff
- increased effectiveness of training
- easier maintenance of documentation
- increased effectiveness of communication within the client's organisation.

Standardising the vocabulary used in a document is no less important than standardising the visual components of the document, components such as non-semantic punctuation, styles and spellings. Indeed, from a communicative point of view, it is far more important. Writers, and managers of writing teams, put considerable effort into creating a house style and style sheets and in ensuring that punctuation, cross-references and lay-out, for instance, are consistent with the set style. But these features, though important, are akin to the body and duco of a car. Those whose interests extend to what is under the bonnet should have a paramount concern with the message rather than with the appearance, and especially with how much effort it takes to communicate what the writers intend to communicate. And the principle of single and distinct denotation is one of the controls that contribute to the communicative efficacy of technical, and indeed any form of, informational writing.

## Addendum 1: The limits of consistency

Minimising the growth of unwanted significance is one reason for writing with consistency as a goal. Another reason is that inconsistency can look sloppy and unprofessional. A perception of sloppiness spreads easily. Readers may pre-judge the quality of the product or your research based on the quality of your documentation or report. That might be an unfair judgement, but it's one that you have no control over once your work has been published.

However, we shouldn't take the push for consistency too far. Inconsistency of vocabulary, for instance, is unlikely to be a problem with common or non-technical synonyms, especially if the synonymity is strong and there is no competing connotation. Hence words such as *but, hence* and *assume* could be swapped with *however, therefore* and *suppose* respectively without causing cognitive dissonance in most mature readers.Usually, inconsistent terminology becomes a problem only with uncommon or technical terms. Thus it would be unwise to swap between *aliphatic hydrocarbons* and *aliphatic compounds*, between *random access memory* and *volatile memory*, and between *propene* and *propylene,* for example.

Certain inconsistencies in design should also be allowed. As a general rule, adhering to a well-designed document template helps the reader associate like structural blocks with like, saves the writer time and thinking, and gives you a better than average chance of generating a document of professional appearance. But there are reasons—good reasons—why you might want to tweak or override a style. And this is something that should only be done once all the drafting and reviewing is finished and you are preparing the document for final release.

Style tweaking is only ever local and usually done when you are considering pagination.[1] Pagination ensures that what

---

1. This should not be read as proscribing *global* style changes so late in the documentation process. Global changes can be made at any time (although they should always be checked). By style tweaking *being local* I mean that the change is made only to one element and not also to all other like elements.

logically should be together on a page is together. For example, there are good readability reasons for ensuring that the one or two rows of a table that spill over onto the next page are brought back to the previous page. To achieve logical pagination, you might need to:

- adjust leading (i.e., line spacing)
- adjust tracking (i.e., spread)
- apply manual page breaks
- delete superfluous words or
- reduce the font size.

Only adjust text if the inconsistency so created is unlikely to be noticed by the reader (such as reducing a font size by 0.5 points or applying a negative spread of 1%).

A good rule of thumb is this: if an inconsistency will not be noticed or, if it is, will not distract the reader from their reading, then don't fuss unduly over it. There are limits to how much homage the God of Consistency deserves.

# Addendum 2: A sample thesaurus

| Keyword | Preferred Term | Notes |
|---|---|---|
| A-party number | | SN: Use to refer to the service number from which a call was made.<br>RT: originating service number, B-party number |
| abnormal *(noun)* | USE abnormal meter record | SN: If describing a meter record that details either a scheduled or inserted meter reading. Note that meter adjustments (i.e., debits and rebates) are not checked for abnormality. |
| abnormal meter reading (1) | | SN: Use when referring to the reading of a meter, not to the record which incorporates that reading.<br>BT: meter reading |
| abnormal meter reading (2) | USE abnormal meter record | SN: If referring to the record which incorporates an abnormal reading |
| abnormal meter record | | SN: Use when referring to a meter record containing a scheduled or inserted meter reading outside the normal range.<br>BT: meter record<br>RT: abnormal meter reading (2) |
| add a case | USE create a case | SN: self-explanatory. Common terminology has users *adding* records to cases. To talk of *adding* a case may suggest to our audiences that-like the error records added to a case-cases must already exist and only need somehow to be selected for use. |
| allocate (1) | USE assign (2) | SN: Use *assign* if talking about supervisors assigning error records to operators. Retain *allocate* if referring to the automatic allocation of records to teams. |
| allocate (2) | | SN: Use if referring to the automatic allocation of records to teams (as occurs via criteria set up in the Rules Table). See (1) in Appendix.<br>RT: assign |
| assign (1) | USE allocate (2) | SN: Use *allocate* if referring to the automatic allocation of records to teams (as occurs via criteria set up in the Rules Table). Retain *assign* if referring to supervisors assigning error records to operators. |
| assign (2) | | SN: Use if referring to supervisors assigning error records to operators. See (1) in Appendix.<br>RT: allocate |
| B-party number | | SN: Use to refer to the service number to which a call has been made.<br>RT: terminating service number, terminating number, destination service number, A-party number |
| batch | USE user-defined set | SN: If meaning a user-created collection of records that is not also a case. The user defines the set of records by providing values for one or more of the selection fields shown on the Criteria Group List screen.<br>RT: group (2), defined group, user-defined group, restricted list, subset |
| bi-annually | USE every six months | SN: If meaning every six months. Avoid using 'bi-' altogether, as some readers will understand it to mean *half* while others will read it as meaning *two*. |

Figure 12.1  An extract from a product-specific thesaurus

## Addendum 3: Editing style sheets

Most editors maintain an editing style sheet as they edit, adding their decisions regarding discretionary vocabulary, punctuation, and the like. By acting as an *aide-mémoire*, these sheets can help writers of whatever persuasion control their vocabulary. Just write your decision in the appropriate box and keep the sheet handy for future reference.

**Editing Style Sheet**

Title: _____ Author: _____ Type of review: _____ Needed by: _____

Special instructions:

| A–B | C–D | E–F | G–H |
|---|---|---|---|
| I–K | L–M | N–O | P–R |
| S–T | U–V | W–Z | Numerics |
| Symbols | Punctuation | Design Issues | Other |

# 13: Font choice and waste

Wouldn't it be a waste if you set out to communicate with readers but there was so much noise in what you had written that communication failed? It would be a waste for you and for your readers. The time spent writing and reading would end up counting for nothing. That distracting noise could arise from many sources: overly complex sentences, ambiguity, vagueness, baffling inconsistencies, emotive language, unfamiliar punctuation and much more. But there is one source of noise that is often overlooked by technical writers, which is odd considering that technical writers these days are expected to attend to it no less than they do to content. That obstacle is poor design. For the contemporary technical writer is a document designer too, and thus rarely able to delegate the tasks of content design and cover design to professional graphic designers.

You might think that content design is merely about aesthetics—but it should be more than that if we want our writing to succeed in its purpose. And it *is* more than that if we want to minimise our environmental footprint.

Firstly, let's consider how design can influence how successful we are in meeting our goal of communicating effortlessly with our intended readers. Clearly, good content design will make reading less tiring, thus improving the chances that you will maximally communicate with your readers. Overly small font size, line spacing that is too tight, pale text colour, inconsistent styling of like paragraphs, uneven word spacing, excessive measure (that is, too many characters per line)—these are just some of the design flaws

First published in *Words*, vol. 1, iss. 2, May 2009

that can make reading unnecessarily difficult and lead, perhaps, to readers disengaging with the text. But one design issue is especially important, and it is one that, these days, seems most ignored by most writers: the choice of font for the body of the document.

## Serif or sans serif?

Between 1982 and 1990, editor and publisher Colin Wheildon conducted a number of experiments to see whether a writer's choice of font could affect readers' comprehension (Wheildon 2005). He gave the nearly 500 participants in his experiments a number of texts to read—some set one way, some another way—and then asked them questions about what they had read. He was particularly interested in the relative communicative effectiveness of *serif* fonts over *sans serif* fonts.

A *serif* is a short line added to the top and bottom of the strokes in traditional typefaces. A *sans serif* font is a font where the characters do not have serifs. In addition, serif fonts mostly use strokes of varying widths while sans serif fonts use strokes of the same widths. Both features can be seen in figure 13.1.

with serifs         without serifs
(sans serif)

Figure 13.1  Serif and sans serif fonts compared

Some commonly used serif fonts are Times New Roman, Perpetua and Palatino. Some commonly used sans serif fonts are Arial, Helvetica and Century Gothic.

Research conducted by the British Medical Council in 1926 suggested that serif fonts are easier to read than sans serif fonts. Wheildon's research backs this up. Half his subjects were asked to read texts with the body text set in a serif font and half were asked to read the same texts with the body text set in a sans serif font. The subjects were then given a series of comprehension tests (being a set of questions relating to the content of each text). Wheildon found that the number of subjects scoring 70% or higher in the comprehension tests was *more than five times higher* if they had read the texts set in a serif font (as shown in table 13.1).

Table 13.1: Comprehension test scores after reading serif and sans serif body text (Wheildon 2005, p. 47)

| | Comprehension test scores | | |
|---|---|---|---|
| | **70–100%** | **40–69%** | **0–39%** |
| | Percentage of subjects with a score in these ranges | | |
| **Serif** | 67% | 19% | 14% |
| **Sans serif** | 12% | 23% | 65% |

Many of Wheildon's subjects in the sans serif group reported difficulty maintaining concentration after reading a dozen or so lines of text, and many reported that they needed to backtrack continually in order to maintain concentration.

It is clear from these results that where retention is the primary goal, texts are best set in a serif font (something that traditional publishers have been aware of for a very long time).

## And for headings?

Wheildon presented his subjects with four styles for headings— serif, sans serif, lower case, upper case—and each was asked if the styles were "easy to read". The results:

- Serif lower case and sans serif lower case were declared easiest to read (by 92% and 90% of subjects respectively).

- Headings set in upper case (serif or sans serif) were not as easy to read as headings set in lower case (with a score of between 3% and 71%).
- Cursive, script or ornamental fonts were declared easy to read by between 11% and 37% (depending on the font).

It seems best, then, to use either serif or sans serif fonts for headings, *but predominantly in lower case*, and to avoid cursive, script or ornamental fonts.

## Capitals for body text?

Wheildon found overwhelmingly that his subjects did not find upper case text (that is capitals) easy to read, whether serif or sans serif.

Table 13.2: Ease of reading upper case text (Wheildon 2005, p. 100)

|                        | Easy to read | Not easy |
|------------------------|:------------:|:--------:|
| Upper case, serif      | 7%           | 93%      |
| Upper case, sans serif | 7%           | 93%      |
| Lower case, serif      | 100%         | 0%       |
| Lower case, sans serif | 22%          | 78%      |

Clearly, large slabs of text should not be set entirely in upper case.

## To bold or not to bold?

Wheildon also presented his subjects with texts set solely in roman or solely in bold (with both texts being set in a serif font). The subjects were again given comprehension tests. The results, given in table 13.3, strikingly show that bold should not be used for large slabs of text.

Table 13.3: Comprehension test scores after reading body text set
roman or bold (Wheildon 2005, p. 51)

| Comprehension test scores | | |
|---|---|---|
| 70–100% | 40–69% | 0–39% |
| Percentage of subjects with a score in these ranges | | |
| **Roman** 70 | 19 | 11 |
| **Bold** 30 | 20 | 50 |

## To slant or not to slant?

Wheildon also presented his subjects with texts set solely in roman or solely in italics (with both texts being set in a serif font). The subjects were again given comprehension tests. The results, given in table 13.4, show that italicising text does not significantly reduce readers' ability to comprehend it.

Table 13.4: Comprehension test scores after reading body text set
roman or italic (Wheildon 2005, p. 51)

| Comprehension test scores | | |
|---|---|---|
| 70–100% | 40–69% | 0–39% |
| Percentage of subjects with a score in these ranges | | |
| **Roman** 67 | 19 | 14 |
| **Italic** 65 | 19 | 16 |

## For on-screen text

The recommendations given above apply to material that is to be printed. For material that is to be read primarily online, Wheildon reverses his recommendation:

"people tend to use different reading modes for screen work, and screen definition is very poor compared with print definition. Indeed, the results [of the research] are almost turned on their head." (Wheildon 2005, p. 21)

But what fonts and styles are best suited for online reading is only part of the equation, *and may even be irrelevant for some types of material*. For in deciding on fonts and styles for online delivery, you need to consider whether the document you are writing is likely to be read online, or merely skimmed and then, perhaps, printed for reading offline.

Given the relatively poor resolution of computer screens — poor relative to the resolution provided by even a low-end laser printer — it is not surprising that many people prefer to read a printed version of what they encounter on screen:

> "the majority of subjects indicated that they preferred reading text from paper." (van de Velde & von Grünau 2003)

> "Survey respondents at all ages (17–61) reported that they usually print digitalized texts before studying them." (Ackerman & Goldsmith 2008)

Research by well-regarded web usability specialist Jakob Nielsen (2008) offers a similar view:

- Seventy-nine per cent of online readers scan the page instead of reading word for word, focusing on headlines, topic sentences and highlighted summaries.
- Online readers are three times more likely than offline readers to read only short paragraphs (one or two sentences long).

Few topics that technical writers write about amount to one or two sentences. It is simply impossible to describe a complex concept, or explain a multi-step process or procedure, in a handful of sentences. So if the document you are writing fits into this category, you might consider adopting fonts and styles more suited to printed material than to on-screen material, *for offline is how most of your readers will read it*.

Why do many people prefer to print out material delivered online? There are at least three reasons. First, reading from the screen is slower — by about 25% — than reading a print-out of the same material (Nielsen 2008).

Second, the online environment is overly distracting. A study reported in *Scientific American* in late 2008 found that:

> "reading online may not be as rewarding—or effective— as the printed word. The reasons: The process involves so much physical manipulation of the computer that it interferes with our ability to focus on and appreciate what we're reading; online text moves up and down the screen and lacks physical dimension, robbing us of a feeling of completeness ... The visual happenings on the screen and your physical interaction with the device is distracting ..."[1]

Third, and no doubt related to the previous point, comprehension of material is less when presented online:

> "Most people, including our survey participants, believe that they learn less efficiently when reading from a computer screen than when reading from paper. We investigated the validity of this belief [and found that] on-screen learners performed worse than on-paper learners under self-regulated study." (Ackerman & Goldsmith 2008)

In fact, this particular study found that comprehension of online material was about 87% of what it would be were it read offline.

Finally, we should note that there is an obligation on writers to write *economically* if it is clear that their readers are likely to print out what they have written. The obligation arises from the moral disbenefit of waste—in this case, the waste both of readers' time and of paper. Clearly, bloated language leads to time theft. The time it takes readers to plough through verbose sentences padded with redundancy and tautology is time that readers could have put to doing the things they—and possibly others—value most.[2] But

---

1. C. Ballantyne, "Online v. print reading: which one makes us smarter?", *Scientific American*, 23 December 2008, http://www.sciam.com. Viewed 15 January 2008. Ballantyne is mostly quoting from Mangen 2008.
2. That such waste is a moral issue is discussed in chapter 2, "Ethics and technical writing" starting page 29.

economical writing can also avoid wasting resources. Suppose you have written a 50-page report and every sentence in it has three superfluous words (a feat that is easily achievable). If we assume that the average length of a sentence is 18 words (James 2007, p. 354), your report effectively contains just 42 pages of substance and eight pages of waste.

You might think that a saving of 12 pages—or any number of pages—is irrelevant in today's digital world. Indeed, the notion of a page means little on the web. A single web page might, if printed, fill half a standard A4 or Letter page, or it might fill ten such pages. Or even twenty. A web page is more like a topic than a physically limited entity. So what does it really mean to save 12 pages on a web page? Nothing, really.

But this overlooks what we noted earlier: when it comes to concentrated reading involving more than four or five paragraphs, most of us will print the document for offline reading. Thus the *printed* length of a document is still relevant in any environmental audit. If a long document destined for the internet will be read offline, then it will be printed on paper. In other words, it will inevitably draw on finite resources. Thus the length of a document is a moral issue. (It is also, of course, a cost issue.) It is worth dropping the page-count of a document from 50 to 38 because the earth is then gouged a little less (leaving more of it for future generations). Is that not a moral issue?

## Paper use and font choice

Factors other than bloated language can cause a document to draw more heavily on natural resources than might otherwise be the case. Font choice is one such factor. To see this, note that the width of most serif font sets is narrower than the width of most sans serif font sets *at the same point size*. Figure 13.2 on page 197 illustrates this. The alphabet appears in six fonts— three serif (Arno Pro, Garamond Premier Pro and Times New Roman) and three sans serif (Arial, Century Gothic and Verdana). The point size is the same in each case.

abcdefghijklmnopqrstuvwxyz

abcdefghijklmnopqrstuvwxyz

abcdefghijklmnopqrstuvwxyz

---

**abcdefghijklmnopqrstuvwxyz**

abcdefghijklmnopqrstuvwxyz

**abcdefghijklmnopqrstuvwxyz**

Figure 13.2 **The width of the printed alphabet varies between fonts**

Notice that the sans serif examples (the bottom three) are significantly longer than the serif examples (the first three). In fact, the shortest serif font illustrated (Arno Pro) is about 75% the length of the longest sans serif font illustrated (Verdana). So a 100-page report set entirely in Verdana could conceivably become just 75 pages if it were set in Arno Pro.

These calculations assume that the characters in the alphabet are used equally throughout the document. This will not be the case in most writing. But even if we restrict the comparison to, say, the 13 most frequently used characters in written English—*e, i, s, a, r, n, t, o, l, c, d, u* and *g*—Arno Pro is still more economical than Arial (the shortest of the three sans serif fonts illustrated): 85 pages for every 100. The same result applies if you compare Garamond Premier Pro with Arial.

It is also true that some sans serif fonts are more economical than some serif fonts. For example, Arial Narrow is more economical than Arno Pro. (Then again, some serif fonts—such as Arno Pro Display—are more economical again.) The point of this section is to emphasise that if most readers print out documents that require concentrated dreading, we risk wasting a valuable resource—paper—if we pay no attention to the fonts we choose.

## Ink use

Most sans serif fonts print darker than serif fonts and thus use more ink. This is largely because the width of the strokes remains the same throughout a sans serif character whereas the widths of the strokes vary within a serif character, sometimes becoming very thin. (Compare the horizontal stroke in the lowercase *e*.) Given that most of us print documents found online if they are more than a few paragraphs long, adopting a serif font for body text may well save on ink as well as on paper.

## Summary

To summarise: injudicious font choice can result in waste in a number of ways. A reader's time is wasted if the use of sans serif fonts forces them to re-read in order to fully understand what they have been reading. The injudicious use of sans serif fonts could also lead to paper wastage (given that most sans serif fonts are greedy for length). And finally, sans serif fonts typically use more ink and toner than serif fonts. In a word, sans serif fonts are less friendly towards readers and the environment than their serif cousins.

## How not to review: Challenging Wheildon

Earlier in this chapter we reported research by Colin Wheildon that shows that an injudicious choice of font for the body of your text can lead to waste. By being prevented from fully comprehending on one pass what they are reading, readers of texts set primarily in a sans serif font have their time wasted. They will have to reread to fully understand the text before them—assuming they have the time to do so or are even aware that they have not fully understood what they have just read.

If this assumption is incorrect, then the writer has also wasted their own time. They set our with the goal of communicating and have inadvertently engineered communication break-down. In a word: their writing has been self-defeating.

The academic literature on the connection between font choice and comprehension is not especially rich. In a 2005 literature review entitled "Which Are More Legible: Serif or Sans Serif Typefaces?" graphic artist Alex Poole states that:

> "There are some high profile studies which claim to show the superiority of serif typefaces (Robinson *et al.*, 1983; Burt, 1959; Wheildon, 1995) but these have been soundly criticised on points of methodology. (Lund, 1997, 1998, 1999)."[3]

What is odd, and somewhat discouraging, about this statement is that in a review of, on Poole's admission, "over 50 empirical studies" he could find only one person critical of Wheildon's work: Ole Lund. Surely if Wheildon's quite extraordinary research had been thoroughly debunked, there would be much more comment in the literature.

Of the three references Poole gives to Lund's work, only one explicitly mentions Wheildon in the title: a three-page review of Wheildon's book *Type & layout: Are you communicating or just making pretty shapes* (Lund 2008, pp. 74–77). This is such an extraordinary review that it is worth dissecting its six short columns column by column. It is a case study in how *not* to write a review and expect it to be seen as carrying any weight.

- Column 1 and half of column 2: Lund gives the history behind Wheildon's publication and a summary of his major result, without analysis or criticism. So far, this is fine, despite the abundance of snigger quotes and the claim that Wheildon's results are "sensational", a claim that is hardly neutral.

---

3.  See http://www.alexpoole.info/academic/literaturereview.html.

- Rest of column 2 and all of column 3: Lund takes Wheildon to task for explaining the results of his research with a "rather dubious appeal to the authority of past research into [the optical phenomenon of] irradiation". But an appeal, dubious or otherwise, to past research is not an argument against the existence of the phenomenon one is trying to explain. Serifs might still provide greater comprehension irrespective of whether irradiation is the reason, just as the earth might still be warming even if solar activity is found not to be the cause.
- Column 4 and most of column 5: Lund quotes a number of the testimonials—from graphic designers, editors and academics—that Wheildon includes in his book and calls it "hilarious" that so many have been included. But even if Wheildon did go over the top in a "persistent appeal for external praise and legitimacy" (as Lund puts it), this is simply not relevant to whether Wheildon's methods and conclusions are sound or unsound.

  The attack on Wheildon, not on his research, starts to get nasty when, of the people Wheildon claims to have provided him with advice during his research—many of whom are or were academics—Lund asserts: "My guess is that it is doubtful that these persons were involved in Wheildon's work in the way that he implies, and that they would not authorise such use of their names". A guess? This is hardly convincing. Indeed, it is a case of *argumentum ad hominem*: attacking the man and not the argument. Even if Wheildon did exaggerate the intellectual influence of others, how is that at all relevant to the issue of font superiority?
- Rest of column 5 and all of column 6: Lund lists two works on information design where Wheildon's results are cited (books by Kempson and Moore, and by Karen Schriver) and he calls them to task for accepting the "blatant outrageousness" of Wheildon's results "more or less uncritically". Uncritically? But where is Lund's logic-based or fact-based criticism?

Lund criticises Wheildon's conclusion but does not provide one skerrick of evidence to contradict it: his own or anyone else's. He talks of Wheildon's "findings" and his "empirical evidence" — snigger quotes included—as if Wheildon never engaged in a study or drew findings from a study. The whole review is a blatant *argumentum ad hominem*, more suited to the opinion pages of a newspaper, not to an academic journal.

Poole gives two other references—again written by Lund— that "soundly criticised on points of methodology" the work of those who thought they had established the superiority of serif fonts. The 1997 reference is to an article titled "Why serifs are (still) important". (Lund 1997) But in this paper, Lund entirely focuses on research reported in 1971 by David Robinson and colleagues.[4] There is not a single mention of Wheildon.

The third reference Poole gives is to Lund's 1999 PhD thesis, entitled *Knowledge Construction in Typography*. The thesis is neither published nor generally available, and thus has not had the benefit of general scrutiny. For that reason, it simply does not deserve to be listed in a literature survey. It may well make some good points against Wheildon, but who can easily tell? I have sent two emails to Lund asking for access to his doctoral material relating to Wheildon. Neither has been answered. It is difficult to comprehend why research that supposedly counters a longstanding publishing practice, and Wheildon's own research, remains unpublished and seemingly guarded by its author. Most academics are more than willing to argue about their views—unless they have abandoned them.

So, *pace* Poole, Wheildon's work has not been "soundly criticised on points of methodology" by Lund. A vitriolic book review lacking all substance does not constitute sound criticism. Nor does a work untested by general peer review. Wheildon's results may well be wrong. But more research— serious research—needs to be done to prove that point (or to back up Wheildon).

---

4.   D. O. Robinson, M. Abbomonte and S. H. Evans, "Why serifs are important: the perception of small print", *Visible Language*, vol. 5, no 4, pp. 353–59.

This is a critical issue. It should interest all technical writers. If we want our readers to maximally comprehend what we write, then font choice may well be a critical decision. If we are writing instructions to assist those in the control room of a nuclear power plant to handle an emergency, our choice of font may well be the difference between life and death.

## How silly can font research get?

Research published in 2010 compared the comprehension and retention of material either set in a fluent font (such as Arial) or a disfluent font (such as Comic Sans). (A disfluent font is one that, by some measure or other, is more difficult to read.) The hypothesis being tested was that "disfluency could lead to improved retention and classroom importance". The researchers found a small but significant improvement in comprehension if students were given texts to read set in Comic Sans or Bodoni—fonts assumed to be difficult to read—rather than Arial (Diemand-Yauman et al. 2010). But given Wheildon's research—see "Serif or sans serif?" on page 190—Diemand-Yauman should also have considered whether Comic Sans or Bodoni are better than serif fonts (such as Arno Pro, Garamond Premier or Times New Roman). But that was left unexplored.

In a second experiment reported in the same article, Diemand-Yauman gave a control group undoctored learning materials and others the same learning materials but reset in Haettentschweiler, Monotype Corsiva or italicised Comic Sans. Again the texts set in disfluent fonts scored better on comprehension tests than the texts that were undoctored. But the fonts used in the control texts are not stated, a staggering omission that must cast doubt on the entire experiment.

The research gained considerable publicity, even making it onto various news services of the Australian Broadcasting Corporation (including *The Science Show*). But the uselessness of this research for improving student retention rates is clear from the fact that only a handful of fonts were compared in the experiment, not one of which was a serif font.

# 14: On wikis and the death of technical writing

The wiki craze has given birth to the idea that the technical writing profession could be in for a bit of a shake-up and possibly a shake-out. If end-users were to write end-user documentation—with a wiki as the collaborative medium— then what role is there for the technical writer? A mere content editor, perhaps—a corrector of misplaced apostrophes and other stylistic perversions perpetrated by untutored writers, nay, "content providers". I heard this view espoused, not unenthusiastically, at a technical communications conference in Wellington in November 2007. The suggestion is that bottom-up collaborative authoring by many heads is better than top-down authoring by one or two heads (that is, by technical writers). The many heads in this case are the users of the product, some of whom would contribute to a product wiki, thus building up a useful body of information about the product likely to exceed, in some way or other, what technical writers could produce.

This might seem an alarming prospect for our profession— but it is just not going to happen. Let me explain why.

The great interest in wikis at present has been spawned by the popularity and success of Wikipedia, the collaborative online encyclopaedia. But the conditions that favour the success of Wikipedia are unlikely to be found in the humdrum world of product documentation. For a start, Wikipedia has tens of thousands of contributors. Moreover, all those

First published in *Southern Communicator*, issue 12, December 2007.

contributors (and every potential contributor) is free to submit a contribution on any topic that takes their fancy. What this means is that contributions to Wikipedia arise from interest and passion. It goes without saying that someone who gives up their spare time to contribute an article to Wikipedia on French history has a passion, or at least a very deep interest, in French history. So Wikipedia is succeeding because an enormous number of contributors are free to write about whatever they are passionate about.

Let's step back from encyclopaedic endeavours and consider the world of product documentation. Imagine a small business owner—someone, say, who runs a restaurant, a nursery, a newsagency or the like—buying an off-the-shelf accounting package. London to a brick: *they are not buying the software out of a passionate interest in accounting.* No, they are buying the software because manual bookkeeping is a soporific drain on the human spirit and yet there is a regulatory requirement that businesses keep a record of their accounts and transactions. Moreover, depending on when they bought the package, the community of users might be as little as 10, or maybe 1,000, maybe 3,000. So, unlike Wikipedia, the community of potential contributors to a wiki that might support users of that accounting package is quite small. And, unlike Wikipedia, the passion to write about the product is just not there. What chance do you think there is that this particular product wiki will accumulate a useful mass of information any time soon?

Moreover, if the customer of this product is a typical small business owner, they will be working 50–60 hours a week. Is this customer likely, in the few free hours on a Sunday morning, to teach themselves the product (which they must do since no documentation was provided by the vendor and the product wiki is either empty or too embryonic to be of much use)? Are they likely to test what they think they have learned, write up their learning (taking special care to write as best they can) and then post it to the product wiki? Airborne pigs come to mind.

The same considerations apply to many of the products technical writers traditionally write documentation for. Take a steam iron, for example. No-one buys a steam iron because they are passionate about ironing. And if they are not passionate about ironing, what chance is there that they will be passionate about writing procedures on how to use the iron? Maybe when the children are in bed, the dishes washed, the carpets vacuumed, the cat fed, the lounge room tidied ... maybe then I'll feel like figuring out how this new iron works and write up some notes for my fellow users. Or not.

The legal ramifications of end-users being given the responsibility for creating end-user documentation are enormous. It is beyond belief that a manufacturer of heavy machinery, or of medical equipment, would release a product to market without comprehensive operating instructions. The risk of litigation should someone be injured or killed as a result of using the machinery or equipment incorrectly is surely not worth the savings to be got by not using technical writers to prepare operating instructions in the first place.

Take an aircraft manufacturer for example. Would Boeing release a new aircraft without operating instructions, in the expectation that pilots would collaboratively compile a wiki on how to fly the plane safely? The company would have to hope that enough pilots lived through the trial-and-error of unassisted learning to compile that wiki. A ghoulish idea, to say the least.

Wikis *can* be useful, but only as an adjunct to traditional technical documentation. (It may well be beneficial for the community of product users if users could add undocumented tips, tricks and work-arounds to a wiki created for that purpose.) But wikis will never replace traditional technical documentation; so those technical writers alarmed at the rise of the wiki need not feel threatened. The noble profession of technical writing will survive wikimania.

# 15: Paragraphing: a vanishing art?

As traditionally defined, a paragraph is a container for one unit of thought and it typically comprises two or more related sentences. Put another way, it expresses one main idea or point and all the sentences in it are related to that main idea or point (Penguin 1993, p. 390).

In informational writing, it is a good idea to begin each paragraph with a *topic sentence*: a sentence that introduces or summarises what the main idea or point in the paragraph is. This is especially useful to those readers—probably the majority of us—who will skim or scan your text looking just for the main ideas or points (Peters 2007, p. 594).

Consider this example:

> "There are two purposes for paragraphs, the one logical and the other practical. The logical purpose is to signal stages in a narrative or argument. The practical one is to relieve the forbidding gloom of a solid page of text." (Hudson 1993, p. 294)

Note the topic sentence at the start of the paragraph. It clearly tells us what the paragraph is about, and each subsequent sentence is clearly related to it and fulfils its promise.

Another good reason for starting each paragraph with a topic sentence is that it helps crystallise your thoughts. It also reduces the likelihood that you will include in the paragraph other matters that may well deserve a paragraph of their own.

First published in *Words*, vol. 2, iss. 1, 2010, pp. 9–14

Still, it is legitimate to start a paragraph in other ways. However, with the skimming reader in mind, it is best always to make it clear in the introductory sentence what purpose the paragraph is serving. For example, suppose you start a paragraph with the sentence "An example is illustrated below". Strictly speaking, this is not a topic sentence, but it does indicate to the reader that the purpose of the paragraph is to illustrate something that was discussed in the previous paragraph. Thus the needs of the skimming reader are met. Another example: "However, there are counter-examples". Again, this is not, strictly speaking, a topic sentence, but it does indicate to the skimming reader that material discussed in the previous paragraph is about to be contradicted.

## Paragraph cohesion and linking words

If a paragraph is a unit of thought expressing one main idea with all the sentences in it related to that main idea, you need to show how the sentences—or more appropriately the ideas in the sentences—are related to one another. When it is clear how all the sentences in a paragraph are related, we say that the paragraph exhibits *cohesion*. The sentences stick together, and reading them provides a fluid, unbroken stream of ideas.

One way of ensuring a fluid, unbroken stream of ideas is to introduce the second and each subsequent sentence with a word or phrase that indicates the link between it and the sentence before it. Common linking words or phrases are *therefore, so, hence, accordingly, it follows then, as a result, in contrast, at the same time, however, further* and *for the same reason*:

> "A number of drugs have shown promise in treating the disease. Doxycycline at 250mg killed 90% of the bacteria. *In contrast*, penicillin at the same dosage killed at best 60%. *However*, penicillin at 500mg killed all the bacteria. *Therefore*, our recommendation …"

A pronoun, such as *it*, can also usefully link one sentence with another, as in the following example:

"Melanoma is a particularly aggressive form of skin cancer. It kills 2000 Australians every year."

Indeed, there is no circumscribed set of words or phrases that will work as linking words. The pool you can choose from is vast, so long as your reader can see the development of your argument or description—that is, see how the distinct ideas in each of your sentences are:

- related to the main topic that the paragraph is about and
- related to each other (through elaboration, extension or qualification).

## Paragraph styling

In traditional publishing, a paragraph is *indented* (or at least every paragraph other than the first paragraph after a heading is indented). This is the style that has been adopted in this book. However, the widespread contemporary practice—at least in commercial writing—is not to indent paragraphs, but to separate them with a blank line (or with extra space above or below the paragraph).

This practice is flawed, and for two main reasons. First, if a paragraph falls at the bottom of a page and its last sentence finishes close to the right margin, there are no obvious cues for the reader to interpret the next sentence—starting at the top of the next page—as the beginning of a new paragraph (if that is what it is). If you have a *space after* or *space before* setting as part of your style definitions, the space is ignored in both Microsoft Word and Adobe FrameMaker. The first line of the next block of texts starts flush against the top margin *regardless of your space settings*. If you use the ENTER key to manually add space between paragraphs—a common though unwise practice— you will get a blank line at the top of the next page. But how can you expect the reader of a print-out of the document to interpret that space as space *in addition to the normal space above the main text frame (that is, in addition to the height of the top margin)*? Readers don't keep in mind the exact height of the top margin so that they can easily detect variances, and most of us wouldn't be able to anyway by eye alone. Moreover, you can't expect the reader to look across at the facing page in order to judge whether both top lines are level. That is simply not how people read. Even so, it would only provide useful information if one, *but not both*, top lines represented the start of a new paragraph. The first-line indent of traditional publishing overcomes this problem.

The first-line indent overcomes another problem inherent in the contemporary practice of using just space to separate paragraphs. This occurs where you have *block text*, that is, text set off from the main body of the paragraph (as in the following example):

> This is an example of block text, text set off from the
> parent paragraph but still part of the parent paragraph.

If space is the only guide a reader has as to when a new paragraph begins, then it will not be obvious to the reader when the text below a piece of block text begins a new paragraph. Note that this particular block of text is not indented, which should make it clear that it is the continuation of the paragraph that began just before the block text—but only because of the convention adopted in this book of indenting new paragraphs (other than the first one after a heading).

You can automate most indenting with appropriate style definitions. Simply specify that the paragraph style to follow each heading style is one without a left indent (say *body_full_out*) and specify that the paragraph style to follow the *body_full_out* style is one with a left indent (say *body_indented*). Finally, specify that the style to follow *body_indented* is *body_indented*.[1]

---

1.  The reason why the first paragraph following a heading is usually not indented is because the purpose of indenting is to ensure that the reader will know when a new paragraph is starting. Clearly, any block of text immediately following a heading is a new paragraph.

## Paragraph length

A paragraph can have any number of sentences. However, it's best to avoid the one-sentence paragraph, increasingly common in newspapers.[2] It is difficult, if not impossible, to develop a thought or idea in just a single sentence. And if you continue the development of that one thought or idea across a number of paragraphs, you run the risk of introducing ambiguity. This is partly because readers will be expecting subsequent paragraphs to be about another idea or thought. It is also because readers have no immediate cues that, say, sentence six is still talking about the topic introduced in sentence one or is talking about a new topic altogether.

One-sentence paragraphs are also inimical to the skimming reader (which is most of us). Where every paragraph is just one sentence, skim reading—the reading of the first sentence in a paragraph—is tantamount to reading the *entire* document. Faced with a document entirely of one-sentence paragraphs, the busy reader is likely to abandon reading before long.

On the other hand, long paragraphs can become overwhelming. They might also amalgamate more than one thought or idea, thereby denying the skimming reader a true glimpse of the main development or argument. So what is an appropriate length?

> "For general purposes, paragraphs from 3 to 8 sentences long are a suitable size for developing discussion, and some publishers recommend an upper limit of 5/6 sentences." (Peters 2007, p. 595)[3]

---

2. The one-sentence command (or step) in a procedure is a notable exception. The practice of separating the command from the result, and from other explanatory notes, generally improves the efficiency with which tasks are done.
3. But note the research in "Part 3: Paragraph length and comprehension" starting on page 142.

## Composing a paragraph

How does one decide what should go into a paragraph and what is best placed in another paragraph? The common definition of a paragraph as a *unit of thought expressing one main idea with all the sentences in it related to that main idea* is a little vague, and thus of limited practical use. We could flesh out the definition to:

> A paragraph is one sentence that expresses the main topic to be discussed, together with other sentences that elaborate, expand on or analyse that main topic.

This is a stronger definition and provides some guidance as to what should go into a paragraph, but it still has practical limitations.

Unfortunately there is no widely accepted formula or algorithm you can apply to determine whether a sentence belongs in one paragraph or another. But if we combine these definitions—with their emphasis on a paragraph necessarily being a container for one main idea or thought—with two facts about how people read, we can pull out of the mixture an algorithm of some practical use. The extra two facts we need to acknowledge are:

- readers mostly find long paragraphs uninviting, and even daunting; they find it easier to process new information if it is provided in chunks
- readers commonly skim or scan, reading the introductory sentence before deciding whether to continue reading the paragraph or skip to the next.

Bearing this in mind, the following approach—illustrated on page 214—might be useful in many forms of declarative or informational writing (although certainly not in novels or short stories). It assumes that you have just written the heading of the section you want to write about and are about to compose the first paragraph after that heading.

## The technique

1. Write a topic sentence, that is, a sentence that introduces or summarises what the main idea is in the paragraph you are setting out to write.
2. Is the sentence conceptually related to the topic suggested by the heading of the section? If not, delete it and repeat from step 1; otherwise continue with step 3.
3. Write another sentence.
4. Is this sentence strongly relevant to the topic sentence?

   If it is not, move it to another paragraph (or delete it) and then repeat from step 3 above. If the sentence is strongly relevant, go to step 5.
5. Are you intending to give the sub-topic introduced with the new sentence in-depth treatment right now (more than two or three sentences)?

   If not, leave the sentence in the paragraph for now and go to step 7; otherwise go to step 6.
6. Are you going to give other sub-topics you intend to add to the paragraph similar in-depth treatment?

   If so, move this sentence to a new paragraph and continue from step 8. In-depth treatment of two or more topics is best spread across several paragraphs rather than crammed into one long paragraph. Remember: readers shun long paragraphs.
7. Is the sentence crucial to the overall description, argument or analysis being expanded in the document (or the current section of the document)?

   If it is crucial, move it to another paragraph and consider making it the topic sentence in that paragraph. (Skimmers only read the first sentence of a paragraph before moving on, and thus if your sentence is essential in whatever description, argument or analysis you are developing, then ideally it should be placed at the start of a paragraph.) If it is not crucial, leave it where it is.
8. If you have more to write about the main topic of the paragraph (as expressed in the topic sentence) then continue from step 3 above. Otherwise review the

paragraph for logical sequence and cohesion and then move on to the next paragraph.

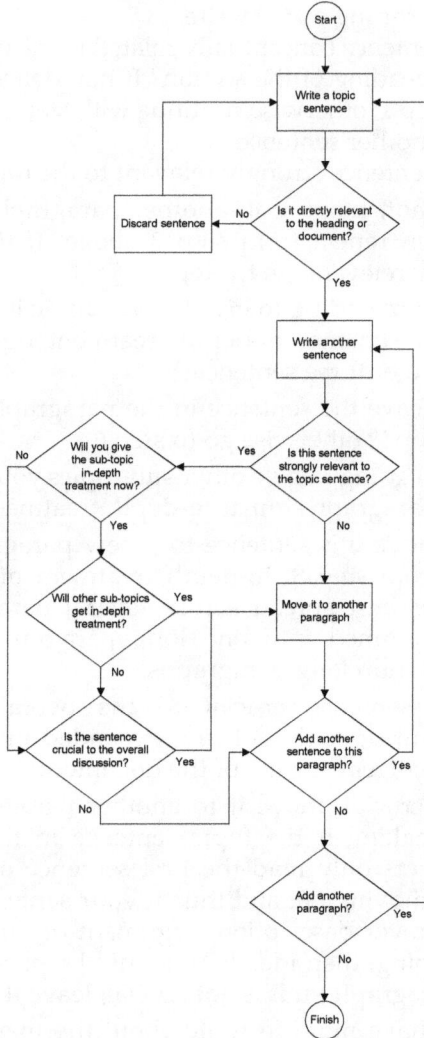

Figure 15.1  Composing a paragraph

## An example

**Setting balloons aloft** [section heading]
One way to set a balloon aloft is to fill it with hydrogen gas.
[*Topic sentence: directly relevant to heading? Yes. So leave the sentence where it is.*]

There are a number of ways of generating hydrogen gas.
[*Strongly relevant to topic? Yes. But this sub-topic is not going to get in-depth treatment, and the sentence is not crucial to the overall development of the information. In other words, if skimmers miss this sentence, there will not be a significant gap in their understanding of the ways of setting a balloon aloft. So leave the sentence in the paragraph.*]

One simple way is to use household caustic soda and aluminium scraps. [*Strongly relevant to topic? Yes. And this sub-topic is going to get in-depth treatment, but no other way of creating hydrogen is going to receive in-depth treatment in this paragraph (otherwise this sentence would be moved to a new paragraph). Further, the sentence is not crucial to the overall development of the information. So leave the sentence in the paragraph.*]

Place three or four tablespoons of caustic soda in a glass bottle with a narrow neck. [*Strongly relevant to topic? Yes. And this sub-topic is not going to get in-depth treatment nor is it crucial to overall development of the passage. So leave the sentence in the paragraph.*]

Another way to set a balloon aloft is to fill it with helium. [*Strongly relevant to topic sentence? No, the topic sentence is about filling a balloon with hydrogen. So this sentence belongs in another paragraph.*]
...

## Another approach

**Setting balloons aloft** [section heading]
You can set a balloon aloft by filling it with a gas that is lighter than air. Two such gases can be readily produced in the amateur chemist's laboratory: hydrogen and helium. [*So far so good. The topic sentence is relevant, and the second sentence is directly elaborating on the topic sentence.*]

To fill a balloon with hydrogen gas … [*The relevance of this sentence to the topic sentence is rather weak. The topic sentence is about how to get balloons to rise, but now we are talking about how to fill a balloon with hydrogen. So this sentence needs to be moved to another paragraph.*]

To fill a balloon with helium gas …[*The relevance of this sentence to the topic sentence is rather weak. The topic sentence is about filling a balloon with hydrogen, but now we are talking about how to fill a balloon with helium. So this sentence needs to be moved to another paragraph.*]

    …

## Good paragraphing: the skimming test

The skimming test is based on the fact that most people skim scientific and technical documents, only reading the paragraphs that are of special interest to them. Skimming is a two-step process: read the first sentence of a paragraph and then decide whether to continue reading or skip to the next paragraph. More often than not most readers will skip at least one paragraph, and usually more. So, if you want to maximise your chance of being fully understood by the maximum number of people who look at your document, make sure that:

- the main point of each paragraph is made in the first sentence and
- all the first sentences, when considered together, provide a satisfactory summary of what you have written.

So the skimming test is this: once you have completed a section of your document, read through it *but just read the first sentence of each paragraph*. As you are reading, ask yourself:

Do all the first sentences, when taken together, adequately portray the gist of the section of the document I have just written? In other words, would they, when strung together (and tweaked with linking words), provide an adequate summary of what I have written?

If the answer is no, then you should reconsider your paragraphing.

For example, if you are presenting an argument and it has four premises, it is unwise to present three of the premises as first sentences and embed the other premise within a paragraph (rather than at the start of a paragraph). The skimming reader will not see the full skeleton of your argument and may think that you have not proved your point even if you have.

Likewise, if you have been describing how to send a balloon aloft by filling it with a gas that is lighter than air and you have described *in roughly equal depth* a number of techniques for producing such a gas, introducing one technique via a first sentence but not the other will not pass the skimming test. For example, the following list of first sentences would not provide an adequate summary if a method of filling a balloon with hydrogen gas was discussed in addition to the method of filling a balloon with helium gas:

> You can set a balloon aloft by filling it with a gas that is lighter than air.
>
> To fill a balloon with helium gas …
>
> You can track the balloon using a simple radio transmitter.
>
> There are a number of ways you can improve your chances of retrieving a spent balloon.

In this case, there needs to be a paragraph with a sentence starting *To fill a balloon with hydrogen gas …* or the like.

Note that while your set of first sentences should provide an *adequate* summary of what you have written, they may not necessarily provide a *good* summary. There may be any number of paragraphs that are providing merely secondary or illustrative material—analogies, parallels, examples and so on—paragraphs that could be deleted without thereby breaking the basic logic of the text. A good summary of that material would not include the gist of the secondary or illustrative material. However, the skimming test is designed only to ensure that all the main points will be picked up by the skimmer, not that the skimmer will only encounter your main points.

# 16: Familiarity vs correctness

Judging by the posts on many online technical writing discussion sites, a good many technical writers believe that language use can be correct or incorrect, and that whatever other attributes technical writing must have for it to be *good* technical writing, perhaps the most important of all is that the language used is correct. Like many in the wider community, these writers consider some usage to be *de rigueur* come what may. To them, it is not a matter of how many people use language in a particular way. Numbers are not important. It is that the language they choose is correct. There are those, for example, who insist that we must—because it is correct—write "the species' distribution" in the singular even if the majority are writing "the species's distribution", the latter practice being what is now recommended in most major contemporary styles guides.[1] Their view is neatly summed up in the following quotation:

> "even if nine-tenths of English speakers were to use a word incorrectly (say, *nauseous* meaning "nauseated" instead of "nauseating"), the remaining tenth would be correct".[2]

For these folk, correctness trumps familiarity. Adopting major-ity usage might make your writing more familiar to your intended audience—since what is most common is, almost by

---

1.  See *Style manual* 2002, p. 85; *Chicago Manual of Style* 2010, p. 353; *New Hart's Rules* 2005, p. 63.
2.  Literary critic Dwight Macdonald, quoted in Steven Pinker, *The Sense of Style*, Allen and Lane, London, 2014, p. 190.

Adapted from Marnell 2015

definition, what must be most familiar. But writers have an obligation (or so many think) not to encourage the incorrect practices of the masses. Such practices are, variously, lazy, corrupting, tempting us into The Precipice of Babel, non-standard or, well, just plain wrong. Thus familiarity can find itself at odds with correctness. This potential conflict between the two is worth exploring in depth, not least because it is a source of language anxiety, and language neurosis, among writers of many stripes.

## Familiarity

If familiarity trumps correctness, then, all other things being equal, we should *always write in ways that will be maximally familiar to our intended audience.* Let's call this the *principle of familiarity.* This principle does have a certain *prima facie* plausibility. If you are writing to communicate—as we mostly are—it would be self-defeating if you use words, idioms, punctuation, symbols and so on that distract or baffle your readers. If you have a deep urge to show off the richness of your vocabulary, then by all means do so—in some form of writing other than informational writing (the writing we are principally concerned with in this book). One does not inform by using language that hinders the informing; nor does one inform efficiently using language that slows the reader. And in the time-poor world we all share, if you don't inform efficiently you might not inform at all, or inform only partly. Thus to achieve your goal of informing, you should write in an audience-centric manner, that is, write for your audience.

Familiarity can be *implicit* or *explicit*. Implicit familiarity is what writing will have if it is written entirely according to the conventions that the audience adopts. Words, idioms, punctuation marks and the like are used exactly as the audience would expect. On the other hand, explicit familiarity is provided by the author to give readers an explanation of what some possibly unfamiliar term or device means. A glossary, for example, gives explicit familiarity. A note at the start of the document alerting readers to special meanings is another case. Technical writers commonly offer explicit familiarity when they include a *documentation conventions* section near the start of a user manual explaining that, say, text set bold is text the user will see on the product and text set in a mono-spaced font is text that needs to be entered. Of course, writing can display a mix of implicit and explicit familiarity.

If you intend to communicate with the least effort on the part of your readers—that is, with reader courtesy in mind—then you should rely as much as you can on *implicit* familiarity. This is the sort of familiarity that does not force readers to break their reading in search of a resolution to whatever might puzzle them. Hence it would be wise to use:

- familiar vocabulary
  Why use *utilise* in a document aimed at a general audience when there is a more familiar, more frequently used equivalent that would be understood by all readers regardless of whether English was their first, second or third language (namely, *use*).[3] Why use *faucet* if you are writing for a British readership?

- familiar meaning
  Engineers and scientists are prone to using common words to mean something other than their common meaning. An example is the word *universe*. Consider the following headline to an article in the popular-science magazine *New Scientist*: "What the universe before ours was like". To most people, the universe is everything there is, was and will be. There simply cannot be a universe before our universe. Another example is *chaos theory*. This is poorly named for it describes difficult-to-predict systems that are nonetheless just as strictly deterministic as any other physical system. There is no chaos in these systems at all, just a difficulty in tying them down to neat formulas.

3. A good proxy measure of the familiarity of a term is its frequency of use in general writing. Useful word-frequency tables are maintained by Professor Mark Davies of Brigham Young University. See www.wordfrequency.info.

- familiar grammar, spelling and punctuation

  There is no point in surprising your readers with hetero-dox syntax, spelling or punctuation, or with syntax, spelling or punctuation that applies to an English other than that used by your intended audience (such us *enroll* for an Australian audience or *red, white, and blue* for a British audience). Unless explicitly flagged in advance, heterodox language will distract readers, breaking their reading and potentially breaking the communication.

There rests the case for the primacy of familiarity in informational writing. There is so much that can get in the way of the laudable goal of communication if we discount the importance of familiarity. However, we can't assume that what is deemed correct will be what is most familiar. So how might a conflict between the two be resolved?

## Correctness

The view that there is correct language use and that it should take precedence over familiarity is called *prescriptivism*.

> "... prescriptivism is the view that one variety of language has an inherently higher value than others, and that this ought to be imposed on the whole of the speech community ... Adherents to this variety are said to speak or write 'correctly'; deviations from it are said to be 'incorrect'." (Crystal 1987, p. 2)

One way to analyse the potential conflict between familiarity and correctness is to consider what *correct* means and then to ask whether, given this meaning, prescriptivism even makes sense. A logical place to start, then, is with dictionary definitions.

The *Oxford English Dictionary* and the *Macquarie Dictionary*, two well-respected dictionaries, give just two definitions of the adjective *correct*. The first is this:

> "correct *adj.*: In accordance with fact, truth, or reason: right." (Oxford)

> "correct – *adjective* 6. conforming to fact or truth; free of error, accurate." (Macquarie)

Both dictionaries give a second definition of *correct* (identically worded):

> "In accordance with an acknowledged or accepted standard; proper."

Given the strident fulminations of many prescriptivists — such as members of the Apostrophe Protection Society, who scrape offending apostrophes from shopkeepers' windows after dark — it is undoubtedly the first sense of correctness they have in mind: in accordance with fact, truth, or reason. For them language use can be correct or incorrect no less than a mathematical equation or scientific claim can be correct or incorrect — or so it seems — and it rankles them when they see supposedly incorrect usage. To them some usage is incorrect come what may. Judgments about usage are thus akin to "2 + 3 = 5" is correct and "Water boils at 50 °C at sea level" is incorrect.

Now if correctness is tied to being factual or true, then to call some usage incorrect — for example, "*So always check your weight after aerobics* is incorrect" — implies that you *know* that this statement is correct (that is, know that it is factual or true). It would be odd indeed to claim that such-and-such a statement is factual or true but that we don't know that is factual or true. Or to claim that we know something but that it might not be factual or true. Truth and knowledge are inseparable. One might back up one's claim by pointing out that this particular instance of language usage is outlawed by a general grammatical rule, namely "It is incorrect to start a sentence with a coordinating conjunction". And from this general rule, the specific judgement — "*So always check your weight after aerobics* is incorrect" — does seem to follow.[4] But this doesn't let the prescriptivist off the hook, for we are entitled to push our demand for proof back one step. For if someone claims that such-and-such a grammatical rule is true or factual, we are entitled to ask how they came by that knowledge (for truth and knowledge are inseparable). In other words, how do we know that "It is incorrect to start a sentence with a coordinating conjunction" is itself correct.

---

4.  However, some linguists call a conjunction at the start of a sentence a *conjunct* rather than a coordinating conjunction. See Peters 2007, p. 168.

# Language and knowledge

To know something is to be *justified* in believing it to be true. As British philosopher A. J. Ayer put it:

> "we do not say that people know things unless they have followed one of the accredited routes to knowledge."
> (Ayer 1956, p.33)

So what are the accredited routes to knowledge? Philosophers have long agreed that there are only two routes to knowledge: by *a priori* means and by *a posteriori* means. *A priori* knowledge is knowledge gained by thought alone; *a posteriori* knowledge is knowledge gained by observation and experimentation.

> "Some statements we can know to be true (or false) only by observation and experiment [such as] *The planet Saturn has rings* …[If] we are to find out whether the statement is true or false we must do some looking around the world (or get someone to do it for us). Statements like these are known as *empirical* or equally, *a posteriori* statements … All other statements are known as *a priori* statements. They are the ones we can know to be true (or false) prior to experience." (Richards 1978, p. 154f.)

## *a priori* knowledge

In an attempt to establish what couldn't be doubted, the French philosopher René Descartes (1596–1650) came up with the famous *cogito erg sum*: I think, therefore I am.[5] Using thought alone, Descartes proved that it is impossible to doubt one's own existence, since the very act of doubting presupposes a doubter. *If I doubt, I exist.* That is a true statement the knowledge of which is gained purely by thought alone. This is an example of *a priori* knowledge. No observation of the outside world was needed to derive it.

---

5.  R. Descartes, *Discourse on Method*, 1637.

Another example: the famous Greek mathematician Euclid (who flourished around 300 BCE) proved that there is an infinite number of prime numbers—that is, numbers divisible only by themselves and by 1, such as 2, 3, 5, 7, 11, 13 and so on. And he did so simply by thinking and applying logic. No-one has ever produced a formula or algorithm that will generate the entire sequence of prime numbers, but, by using thought and logic alone, Euclid proved that there must be an infinite number of them.[6]

Could the truth (or falsity) of the claim *All sentences that begin with a coordinating conjunction are not in accordance with fact, truth or reason* be gained by thought or logic alone along the lines of Descartes and Euclid? For *a priori* deduction to succeed, it must be based on either an axiom or a definition. (An axiom is a proposition considered to be self-evident, such as *Two points in two-dimensional space can always be joined by a straight line*.) An axiom or definition is needed to halt what would otherwise be an infinite regress of subsidiary justifications. One or the other is a necessary starting point in any deduction. Euclid makes use of both: much of his geometry—the geometry still taught in schools—is based on axioms, and his proof of the infinity of prime numbers is based on a definition (of *prime number*). What axiom or definition might conceivably compel us to accept that all sentences that begin with a coordinating conjunction are not in accordance with fact, truth or reason?

For a start, we are on safe grounds in ruling out axioms as the bedrock for grammatical knowledge. If the so-called rules of grammar were axiomatic—that is self-evident truths—there would be no disputation among linguists and grammarians (just as there is no disputation among mathematicians as to whether the angles in a two-dimensional triangle sum to 180°). But there is disputation. Moreover, if self-evident, the teaching of grammar could then be simplified by mirroring the teaching of mathematics: axiom–deduction–fact. The fact that

---

6. Euclid, *Elements*, book 9, proposition 20.

grammar teaching isn't like this should indicate that teachers—folk who presumably know a thing or two about grammar—don't consider grammatical rules to be self-evident. That leaves us with definition. But it should be clear that so-called grammatical knowledge is not based on definitions. Nothing about "Never split an infinitive", "Never start a sentence with a coordinating conjunction", "Never dangle a participle" and so on makes it true by definition. Indeed, if grammatical knowledge could be based on definitions, the rules of correct usage would pop out of syllogisms in much the same way as philosophical statements do. But not even the prescriptivists accept that grammatical knowledge is akin to philosophical knowledge.

## a posteriori knowledge

The only other way to gain knowledge is by observation or experiment. For example, we know that nitrogen is odourless because, by observing pure samples of it, we do not detect an odour. We know that water boils at 100°C at sea level because many people have done the experiment and got the same result. Most of what we know comes to us by observation and experiment, that is, by *a posteriori* means. Every scientific discovery provides us with *a posteriori* knowledge: our knowledge of medicine, of chemistry, of astronomy.

It should be clear that we don't arrive at grammatical knowledge—assuming that there can be such a thing—by *direct observation*. We do not, for instance, directly observe the truth of "It is incorrect to start a sentence with a coordinating conjunction" as we might "Nitrogen is odourless". Nor do arrive at grammatical knowledge by *induction*, that is, by conducting numerous experiments with sentences that begin with a coordinating conjunction and observing that in all cases they exhibit incorrectness, thus enabling us to infer that all such instances must be incorrect (just as we infer that water always boils at 100°C at sea level). All this should be obvious from the fact that we don't teach grammar as we do science.

This forces on us an incontrovertible conclusion: since knowledge can only be gained by *a priori* means or *a posteriori* means (that is, by thinking or by observation) and neither can yield rules of language use, the rules of language use cannot be proper subjects of knowledge. In other words, they cannot be factual or true.[7]

## Correctness and standards

In so far as it equates grammatical correctness with grammatical knowledge, the first dictionary definition of *correct*—"in accordance with fact, truth, or reason"—is clearly untenable. Knowledge cannot be invoked to justify a claim about the correctness or otherwise of language usage. Let's now consider the second dictionary definition:

> "In accordance with an acknowledged or accepted standard; proper."

Using our earlier example, we can imagine this definition of *correct* being put to use in the following way:

[1] For a practice to be correct, it must be in accordance with an acknowledged or accepted standard.

[2] If a practice is not in accordance with an acknowledged or accepted standard, it is incorrect.

[3] The practice of beginning a sentence with a coordinating conjunction is not in accordance with an acknowledged or accepted standard.

[4] *So always check your weight after aerobics* begins with a coordinating conjunction.

[C] Therefore *So always check your weight after aerobics* is incorrect.

This argument is valid: the conclusion [C] logically follows from the four premises. However, we need to consider whether the argument is also sound. In other words, are the premises true?

---

7.   These arguments are fleshed out in Marnell 2015, chapter 2.

Let's begin by considering the meaning of *standard*, the pivotal word in premise [1]:

> "standard *noun* anything taken by general consent as a basis of comparison; an approved model" (*Macquarie Dictionary*)

From this definition, it is possible to discern two types of standard: *explicit* and *implicit*. Explicit standards are invented, often formalised and sometimes legislated. There is an acknowledged or accepted body that creates or maintains them. For example, there is an explicit standard governing the game of chess and a body authorised to maintain it (authorised, perhaps, by the considered decisions of elected representatives from numerous chess-players' associations worldwide). This body is the World Chess Federation (also known as FIDE, which stands for *Fédération Internationale des Échecs*). Chess players agree to play by this standard — that is, according to the rules of chess — and will modify how they play the game should FIDE change the standard (that is, change the rules). *Implicit* standards lie behind behaviours that, although not directed by an explicit standard, are done widely enough to be considered routine and even expected. We talk of it being *standard* practice to shake hands when meeting a new acquaintance, or waving when farewelling someone, or buying a present for one's partner on their birthday and so on. There are no authoritative bodies that direct us to do these things. If we feel compelled to do them, the compulsion is private. We might do it out of politeness, a desire to fit in or a wish not to be seen as a curmudgeon. But if we fail, say, to shake the hand of someone we are being introduced to, we are considered to be acting in a *non-standard* way. We have gone against a custom.

Now if language use follows an explicit standard, there must be some body that has authority over the standard, some body authorised to set or maintain usage (just as FIDE has the authority to set and maintain the rules that comprise the standard to be adopted by chess players). But there is no such body for the English language. The so-called rules of English are not set and controlled by anyone. (Indeed any suggestion that they should be is unlikely to gain favour throughout the English-speaking world. Where might the authority derive to force an English-speaking country to give up its variant of English in favour of some other variant, or some artificial variant concocted by a committee?) If there is no body that has authority over the use of the English language then we must conclude that its use—in so far as it has more than just private currency—follows *implicit* standards, with one such standard for each variant of English (American English, Yorkshire English, Australian English and so on).

Given that language use is not directed by an explicit standard but follows (or sets) an implicit standard, the definition of *correct* we are here exploring can be recast as:

> In accordance with an acknowledged or accepted implicit standard; proper.

We noted earlier that standards arise "by general consent". Note now the definition of *convention*:

> "convention *noun* a rule, method, or practice established by general consent or usage" (*Macquarie Dictionary*)

Since an implicit standard is obviously a convention, we can recast our definition of *correct* in terms of conventions:

> In accordance with an acknowledged or accepted convention; proper.

Conventions are either *artificial* or *natural*. Artificial conventions are practices that result from following *explicit* standards. Explicit standards are concocted to serve a special purpose. They do not arise naturally. Thus the conventions that result from people following them are artificial. Moving a bishop diagonally in chess is conventional. It matches a general practice. Moreover, that general practice happens to follow what is set out in an explicit standard. There is nothing natural about chess. It did not arise out of any force of nature. It is a game *Homo sapiens* invented. In other words, it is an artificial construct. On the other hand, natural conventions are not concocted, but arise unaided. They are not the *product* of any standard. Rather it is the widespread adoption of a practice that makes it a natural convention and thus a standard. A natural convention is the practice of waving to farewell someone. Another is a natural language, such as English.

Given that language use follows an implicit standard and whatever follows an implicit standard is a natural convention, we can recast our definition of *correct* as:

> In accordance with an acknowledged or accepted natural convention; proper.

But such a definition spawns some counter-intuitive implications. First, there have been many widely accepted natural conventions the breach of which could not conceivably be incorrect. Before the 1960s, there was a strong natural convention in Australia—and in much of the Western world—that saw women adopt their fiancé's surname after marriage. That was the convention at the time: widely accepted, indeed expected. But those women who refused to adopt their fiancé's surname didn't do something that could be considered *incorrect*. They simply did something that, at the time, was unconventional. Similarly, there was once a strong convention that a woman didn't engage in paid employment but maintained a household for her husband and family. Was it therefore *incorrect* for women to work outside the home? (And were those employers who offered them a job equally complicit?) Likewise, it was not that long ago that a widely accepted natural convention saw most people in Western countries attend church every week. Were the atheists who refused to do so doing something incorrect? Similarly, it was once conventional for men to wear black-tie at dinner. Were those who decided to don lounge suits instead doing something incorrect? Hardly. Hence any definition that ties correctness to following a natural convention—and, by extension, incorrectness to flouting a natural convention—appears far too general. It strips the word of its strongly pejorative connotation. To call a behaviour *incorrect* implies that it should be discouraged. But why should a woman be discouraged from keeping her own surname after marriage just because the majority happen to change their name? Why should I be discouraged from wearing a lounge suit to dinner just because the majority of my contemporaries prefer black-tie?

Moreover, an action might be in accordance with an acknowledged or accepted standard and yet be morally repugnant. Slavery, infibulation, genital mutilation and racial cleansing have all, at some time or another, been standard or conventional in some societies. Are we to say that those actions must therefore have been correct for members of those societies?

The definition also implies that most of what we do is incorrect, for most of what we do is not covered by acknowledged or accepted conventions. For example, it implies that collecting orange peel is incorrect. It is not in accordance with an acknowledged or accepted standard precisely because there is no standard governing the collection of orange peel. So am I to be discouraged from collecting orange peel if that is my inclination? To argue so would be odd.

Attempts to improve language further strain a convention-based definition of correctness. At one time there were no spaces between words (a form of writing known as *scriptio continua*). In other words, there was no convention to place spaces between words (or, conversely, a strong convention of not placing spaces between words). The practice impeded reading, not to mention generated unnecessary ambiguity. When spaces first began to appear between words, the practice would have been unconventional. But it would be somewhat odd to call an unconventional practice incorrect if it avoids ambiguity and thereby improves the language. Must every improvement necessarily be incorrect when it is introduced? Given the strong pejorative connotation of *incorrect* — that the behaviour to which it is applied should be discouraged — it is difficult to accept that a practice could be incorrect if it is beneficial. Why discourage what is obviously good?

Tying correctness to following a natural convention also strips the concept of its connotation of universality. A practice only becomes a convention when a critical mass of people adopts it. So at time $t_1$ a practice might not be conventional and at time $t_2$ it is conventional. Perhaps little more than a day separates the two, the day on which one extra person adopts the practice and the balance dips to the other side. Thus a person who had adopted the practice before $t_2$ could be acting incorrectly one day but correctly the next. This sort of relativism swims against a strong connotation of *correct* and *incorrect*: the connotation of universality. An act might be *unlawful* one day but *lawful* the next—laws being at the discretion of parliaments—but to claim that an act can be *incorrect* one day and *correct* the next (or vice versa) strikes a dissonant chord.

It might be retorted that we have too loosely interpreted the meaning of *correct* and *incorrect* given in the dictionaries. Our argument that these definitions imply that collecting orange peel is incorrect, or that a practice can be incorrect one day but correct the next, assumes that a practice can be incorrect *in the absence of a standard*. Perhaps we can avoid some of the problems we have adduced by qualifying the dictionary definition so that it applies only when a standard already exists. Thus collecting orange peel is neither correct nor incorrect, for no standard pertaining to the collection of orange peel exists. There is a parallel here with morality. Some actions are *amoral*, that is, neither moral (those that are to be encouraged) nor immoral (those that are to be discouraged). Scratching my nose is amoral, as is smiling at a neighbour. Just as the concepts *moral* and *immoral* are contrary rather than contradictory, so too, perhaps, are the concepts *correct* and *incorrect*. To coin a word, there could be *acorrect* actions: those that are neither correct nor incorrect. Allowing this recategorisation overcomes the problem of being saddled with definitions that imply that seemingly innocuous actions—and every novel action—are necessarily incorrect.

But this re-interpretation doesn't overcome all the problems we have pointed out. Apart from it being a re-interpretation— and thus going beyond the findings of lexicographers —it would still declare many breaches of natural conventions incorrect when they clearly are not. There *was* an existing natural convention at the time when women decided that they didn't need to change their surname after marriage. By not doing so, many women would have done what our new definition implies is incorrect. *Scriptio continua* was a natural convention when the irish monks first put spaces between English words, and thus it was incorrect of them to improve the English language by doing so. The new definition also runs counter to the universality connotation of the words *correct* and *incorrect*. Just as it makes no sense to call a practice incorrect at time $t_1$ but correct at $t_2$, it makes no sense to call a practice acorrect at $t_1$ but correct at $t_2$.

These considerations compel us to reject the second dictionary definition of *correct*, namely, in accordance with an acknowledged or accepted standard. For clearly it is not a cohesive or even informative definition (whether or not we add the qualifier that the standard must already exist). Its sieve is too tightly woven to capture only what is incorrect. That, together with the illogical implications that follow from it, suggest that its use must be fuzzy, inconsistent and poorly thought-out. It might be innocuous to call breaches of *artificial* conventions incorrect (such as moving a bishop vertically in a game of chess). But natural conventions are sand-bagged against such attributions. To call a breach of a natural convention—such as failing to take off one's hat in a church or failing to end a sentence with a full stop—incorrect is to make what philosophers call a *category mistake*. It is akin to attributing greenness to honesty or polarity to smiling.[8]

There is another way we can show that language use cannot be correct or incorrect. Consider any usage that a prescriptivist might call incorrect. For example, suppose *nite* is declared to be an incorrect spelling of *night*. What, then, are we make of this passage from a Charles Dickens novel:

---

8.   "[A category mistake represents facts] as if they belonged to one logical type or category (or range of types or categories), when they actually belong to another." Gilbert Ryle, *The Concept of Mind*, Hutchinson, London, 1949, p. 17.

"God forbid as I, that ha' known, and had'n experience
o' these men aw my life—I, that ha' ett'n an' droonken
wi' 'em, an' seet'n wi' 'em, and toil'n wi' 'em, and oov'n
'em, should fail fur to stan by 'em wi' the truth, let 'em
ha' doon to me what they may!"[9]

It would be odd to accuse Dickens of *incorrect* spelling in
this passage. The spellings might not match those adopted
elsewhere in the novel (or taught in the schools of nineteenth-
century England). But that is what Dickens intended. Now if it
was not *incorrect* of Dickens to intentionally introduce so many
non-standard spellings, why is my equally intentional non-
standard spelling of *night* as *nite* incorrect? One might retort
that Dickens was trying to achieve some special effect. Well,
yes; but so might I. I might, for example, be actively involved
in a campaign to reform English spellings, inspired perhaps by
Harry Lindgren or George Bernard Shaw. Surely correctness
cannot apply to some writers but not others. If it could, we
would have to strip the word of its universality connotation.
We would also have to address the thorny issue of why some
authors are allowed artistic licence and others are not.

But what if I intended to spell *night* according to its
conventional spelling and spelt it as *nite* instead? Doesn't this
change matters? Can't writing *nite* while intending to be
conventional be called incorrect? The short answer is: no.
Suppose I didn't intend to over-cook the asparagus, but I did.
Was my cooking *incorrect*? No. That would be another example
of a category mistake. So why must my *nite* be incorrect if I
intended to write *night*? I wanted to cook the asparagus
lightly/spell *night* as it is commonly spelt, but I failed. If it's not
incorrect in the first case, why is it incorrect in the second? At
most I have made a *mistake*, and to be mistaken is not
necessarily to be incorrect. *Mistaken* carries none of the
pejorative energy of *incorrect*. To say *I mistook Peter for Paul* is
not to imply that I behaved incorrectly. Another example:
suppose I intend to wear black-tie to a dinner of The Oxford

---

9.    C. Dickens, *Hard Times*, Chapman & Hall, London, 1911, p. 164.

Society (where black-tie is the longstanding unwritten conventional dress code). Suppose further that I wrongly think that black-tie just means wearing a tie that is uniformly black. I arrive at the dinner wearing jeans and a black tie. It would, I suggest, be unduly stretching the meaning of words to call my dress that evening *incorrect*. I have simply made a mistake.

Further, if unintentional language mistakes could be labelled incorrect, we would have to allow that a particular instance of language use could be both correct and incorrect. To see this, suppose that Mary and Paul both write "Mat the cat sat on the". If Mary knew that the sentence didn't follow conventional English syntax and Paul thought it did, we would have to say, of the very same sentence, that it is both incorrect (Paul's version) and not incorrect (Mary's version). That is an impossible position to maintain. But if we allow *intention* into the argument as a mitigating condition—as we must if we are not to criticise Dickens for his spelling—then that is the position we would be forced to adopt. Since it is clearly untenable—it involves a logical contradiction—so too is the claim that unintentional language mistakes are incorrect.

To sum up: if I *intentionally* use unconventional English, what I have done is not incorrect. (That is the Dickens defence.) If I *unintentionally* use unconventional English, the most that can be said is that I have made a mistake. Whatever error there might be is with me, not with what I have written. I intended to write conventionally but failed to do so.

*

We have considered the only definitions of the adjective *correct* provided in two respected dictionaries: *Oxford English Dictionary* and *Macquarie Dictionary*. The first definition ties correctness to fact, truth or reason. This definition would, if applied to the so-called laws of language, make them subject to the same rigorous challenges that can be put to the laws of mathematics and science: verification or falsification. For them to be knowledge, their truth must be ascertainable. We found, however, that the truth of grammatical laws cannot be ascertained. There are simply no accredited epistemological paths open to us to verify or falsify them. Hence they must be beyond the scope of that particular definition of *correct*. The second definition ties correctness to being in accordance with an acknowledged or accepted standard. In the case of language, this definition is equivalent to being in accordance with a natural convention. But breaches of natural conventions are not typically considered incorrect. This second definition, then, is clearly at odds with how the concept of correctness is generally understood. It is too deformed or too fragile to be of any use in the debate about whether language use can be correct or incorrect.

To sum up: in matters of language use, the apparent conflict between familiarity and correctness is a false construct. It cannot occur, and the reason is simple: incorrectness cannot occur. It is best for writers to stick with the principle of familiarity discussed earlier in this paper. Writing to rules that contradict majority usage is likely to yield self-defeating writing for the sake of a goal that has no philosophical merit.

*

Our approach in this chapter has been to dissect dictionary definitions of *correct*. Prescriptivists won't complain about this, as they are forever chiding others for using words in ways not sanctioned by dictionaries. But, as we have shown, dictionaries cannot always be relied on to prove points and resolve arguments. This is not the fault of lexicographers, the folk who compile dictionaries. With few exceptions in the field of dictionary publishing, lexicographers record *actual* usage, and actual usage is often coloured by people's *beliefs*. And dubious beliefs can lead to dubious definitions (and the second definition of *correct* we dissected is clearly dubious).

The fact that dictionaries are not infallible guides to clear thinking does not mean that our approach in this chapter has been misguided. If we are to avoid straw-men debates, we must use concepts that have currency. Dictionaries attempt to pin down those concepts and their meanings. That is their primary purpose, however vague the concepts. Thus they make the best starting point in any debate.

To encounter generally accepted concepts that are poorly thought out is part and parcel of human enquiry. The history of thinking — whatever the discipline — is littered with concepts once thought valuable but subsequently found wanting. Think of *aether*, *phlogiston* and *forms* (the latter of the Platonic variety). Dictionaries will try to make sense of such concepts. And they may well fail (as two have with the concept of *correctness*). But analysis must start somewhere. And if it can't start with common meanings, as sketched out in dictionaries, where can it start? It would be pointless starting with the interpretations of eccentrics whose private definitions are alien to all but an insignificant few. Thus the invaluable role of dictionaries — warts and all — in attempts to clarify our ideas and resolve our disputes.

# A: Analysis of survey results

In 2009 I created a web-based survey designed to gather the thoughts of practising technical writers worldwide on what their profession should be called. The preferred titles are listed on page 22.

The survey results were also analysed by country, employment type (contractor or staff) and years of experience. These analyses are described below.

## Suggested titles by country (and in order of preference)

### Australia (45 responses)

- technical writer, 20 (44.5%)
- technical communicator, 10 (22%)
- information designer, 5 (11%)
- technical author, 3 (6.5%)
- instructional writer, 2 (4.5%)
- content developer, 1 (2%)
- document specialist, 1 (2%)
- information specialist, 1 (2%)
- information developer, 1 (2%)
- user assistance and language expert, 1 (2%)

### Canada (6 responses)

- instructional writer, 2 (33%)
- technical writer, 2 (33%)
- information designer, 1 (17%)
- technical communicator, 1 (17%)

## India (10 responses)

- information designer, 3 (30%)
- technical communicator, 3 (30%)
- content developer, 1 (10%)
- content specialist, 1 (10%)
- information engineer, 1 (10%)
- technical writer, 1 (10%)

## New Zealand (35 responses)

- technical communicator, 8 (23%)
- technical writer, 7 (20%)
- documentation developer, 6 (17%)
- information designer, 6 (17%)
- documentation specialist, 3 (8.5%)
- technical author, 2 (6%)
- content writer, 1 (3%)
- documenter, 1 (3%)
- information architect, 1 (3%)

## United Kingdom (17 responses)

- information designer, 4 (23.5%)
- technical writer, 4 (23.5%)
- technical communicator, 3 (17.5%)
- technical author, 2 (12%)
- communications specialist, 1 (6%)
- documenter, 1 (6%)
- information developer, 1 (6%)
- information technician, 1 (6%)

## United States (47 responses)

- technical writer, 15 (32%)
- technical communicator, 11 (23.5%)
- technical author, 6 (13%)
- information developer, 5 (11%)
- information designer, 4 (8.5%)
- content delivery architect, 1 (2%)
- content developer, 1 (2%)

- content provider, 1 (2%)
- developmental editor, 1 (2%)
- technical communication professional, 1 (2%)
- user support designer, 1 (2%)

**Other countries (5 responses)**

- documentation specialist, 1 (20%)
- information designer, 1 (20%)
- technical communicator, 1 (20%)
- technical journalist, 1 (20%)
- technical writer, 1 (20%)

## Contractor preferences

Thirty per cent of respondents (50) are contractors. Their preferences are:

- technical writer, 19 (38%)
- technical author, 8, (16%)
- technical communicator, 5 (10%)
- instructional writer, 4 (8%)
- information designer, 3 (6%)
- content developer, 2 (4%)
- documentation developer, 2 (4%)
- [other, < 2], 7 (14%)

## Employee preferences

Seventy per cent of respondents (115) are employees. Their preferences were quite different from those of contractors:

- technical communicator, 32 (28%)
- technical writer, 31 (27%)
- information designer, 21 (18%)
- information developer, 6 (5%)
- technical author, 5 (4%)
- documentation developer, 4 (3.5%)
- documentation specialist, 3 (2.5%)
- documenter, 2 (2%)
- [other, < 2], 11 (10%)

## Preference by experience

### 0–4 years (32 responses)

- information designer, 10 (31%)
- technical writer, 8 (25%)
- technical communicator, 4 (12.5%)
- documentation developer, 2 (6%)
- [other, < 2], 8 (25%)

### 5–9 years (39 responses)

- technical communicator, 13 (33%)
- technical writer, 10 (25.5%)
- technical author, 4 (10%)
- information designer, 3 (7.5%)
- [other, < 2], 9 (23%)

### 10–14 years (29 responses)

- technical communicator, 10 (33.4%)
- technical writer, 7 (24%)
- documentation developer, 3 (10%)
- information designer, 2 (7%)
- information developer, 2 (7%)
- technical author, 2 (7%)
- [other, < 2], 3 (10%)

### 15–19 years (25 responses)

- technical writer, 12 (48%)
- information designer, 6 (24%)
- technical communicator, 4 (16%)
- [other, < 2], 3 (12%)

### 20+ years (40 responses)

- technical writer, 13 (32.5%)
- technical communicator, 6 (15%)
- technical author, 5 (12.5%)
- content developer, 3 (7.5%)
- information developer, 3 (7.5%)

- information designer, 2 (5%)
- instructional writer, 2 (5%)
- [other, < 2], 6 (15%)

# Bibliography

Ackerman R. and Goldsmith M. 2008, "Learning Directly From Screen? Oh-No, I Must Print It!: Metacognitive Analysis of Digitally Presented Text Learning", *Proceedings of the Chais Conference on Instructional Technologies Research*. Available at http://telem-pub.openu.ac.il. Viewed 10 April 2008.

Ayer A. J. 1956, *The Problem of Knowledge*, Penguin, Harmondsworth.

Baddeley A. 1987, *Working Memory*, Oxford University Press, Oxford.

—— 2007, *Working Memory, Thought, and Action*, Oxford University Press, Oxford.

Bradley F. H. 1876, Ethical Studies, Oxford University Press, London.

Chall J. S. 1958, *Readability: An Appraisal of Research and Application*, Ohio State University Press, Columbus.

Cheek A. 2010, "Defining Plain Language", *Clarity: Journal of the International Association Promoting Plain Legal Language*, no. 64, November 2010.

*Chicago Manual of Style* 2010, University of Chicago Press, Chicago,16th edn.

Cowan N. 2000, "The Magical Number 4 in Short-term Memory: A Reconsideration of Mental Storage Capacity", *Behavioral and Brain Sciences*, vol. 24.

—— 2010, "The Magical Mystery Four: How is Working Memory Capacity Limited, and Why?", *Current Directions in Psychological Science*, vol. 19, iss. 1.

Crystal D. 1987, *The Cambridge Encyclopedia of Language,* Cambridge University Press, Cambridge.

Doiemand-Yauman C., Oppenheimer D. M. & Vaughan E. B. 2010, "Fortune Favors the Bold and the Italicized: Effects of Disfluency on Educational Outcomes", *Cognition,* DOI: 10.1016/j.cognition.2010.09.012.

DuBay W. H. 2007, *Smart Language: Readers, Readability, and the Grading of Text,* Impact Information, Costa Mesa, CA.

Finegan E., Besnier N., Blair D. & Collins P. 1992, *Language: Its Structure and Use,* Harcourt Brace Jovanovich, Sydney

Flesch R. 1948, "A New Readability Yardstick", *Journal of Applied Psychology,* vol. 32, iss. 3, pp. 221– 233.

Fowler H. W. 1965, *A Dictionary of Modern English Usage,* Oxford University Press, New York, 2nd edn.

Gorenflo D. W.& McConnell J. V. 1991, "The Most Frequently Cited Journal Articles and Authors in Introductory Psychology Textbooks", *Teaching of Psychology,* vol. 18., no. 1.

Gould S. J. 1996, *The Mismeasure of Man,* Norton, New York.

Hargis G., Carey M., Fernandez A. K., Hughes P., Longo D., Rouiller S. & Wilde E. 2004, *Developing Quality Technical Information: A Handbook for Writers and Editors,* Prentice Hall, Upper Saddle River.

Harris S. 2010, *The Moral Landscape,* Free Press, New York.

Haydon L. M. 1995, *The Complete Guide to Writing and Producing Technical Manuals,* Wiley-Interscience.

Horn R. E., 1992, *Developing Procedures, Policies & Documentation,* Info-Map, Waltham.

Hudson N. 1993, *Modern Australian Usage,* Oxford University Press, Melbourne.

INTECOM (The International Council for Technical Communication) 2003, *Guidelines for Writing English-Language Technical Documentation for an International Audience*. See http://www.tekom.de/upload/alg/INTECOM_Guidelines.pdf.

ISO (International Standards Organization) 2008, *ISO/IEC 26514, Systems and software engineering—Requirements for Designers and Developers of User Documentation*.

James N. 2007, *Writing at Work*, Allen & Unwin, Crows Nest.

Kintsch W., Kozminsky E., Streby J., McKoon G. & Keenan J. M. 1975, "Comprehension and Recall of Text as a Function of Content Variables", *Journal of Verbal Behavior and Verbal Learning*, vol. 14, iss. 2.

Kintsch W. & Rawson K. 2005, "Comprehension", in M. J. Snowling & C. Hulme, *The Science of Reading: A Handbook*, Blackwell, Oxford.

Kuhn T. 1962, *The Structure of Scientific Revolutions*, University of Chicago Press, Chicago.

Lund O. 1997, "Why Serifs are (still) Important", *Typography Papers*, iss. 2, pp. 91–104.

—— 1998, review in *Information Design Journal*, 1998, vol. 9, iss. 1, pp. 74–77.

Mangen A. 2008, "Hypertext Fiction Reading: Haptics and Immersion", *Journal of Research in Reading*, vol. 31, iss. 4, pp. 404–419.

Markel M., Vaccaro M. & Hewett T. 1992, "Effects of Paragraph Length on Attitudes Toward Technical Writing", *Technical Communication*, Society for Technical Communication, vol. 39, iss. 3, pp. 454–6.

Marnell G. 2015, *Correct English: Reality or Myth?*, Burdock Books, Brighton.

Miller G. A. 1956, "The Magical Number Seven, Plus or Minus Two: Some Limits on our Capacity for Processing Information", *The Psychological Review*, vol. 63, no. 2.

*New Hart's Rules: The Handbook of Style for Writers and Editors* 2005, Oxford University Press, Oxford.

Nielsen J. 2008, "Why Web Users Scan instead of Read". Available at http://www.useit.com/alertbox/ whyscanning.html. Viewed 9 April 2008.

O'Keefe S. 20018, "XML, Growing Up Fast", *INTECOM*, July/ August 2008.

Parfitt D 1984, *Reasons and Persons*, Clarendon, Oxford.

*Penguin Working Words: An Australian Guide to Modern English Usage* 1993, Penguin, Ringwood.

Peters P. ed. 1989, *The Macquarie Student Writers Guide*, Jacaranda, Milton.

—— 2007, *The Cambridge Guide to Australian English Usage*, Cambridge University Press, Cambridge (UK).

Peterson L. R. & Peterson, M. J. 1959, "Short-term Retention of Individual Verbal Items", *Journal of Experimental Psychology*, vol. 88, no. 3.

Pollack I. 1952, "The Information of Elementary Auditory Displays", *Journal of the Acoustical Society of America*, vol. 24., no. 6.

Quirk R., Greenbaum S., Leech G. & Svartnik J. 1972, *A Grammar of Contemporary English*, Longman, London.

Richards T. J. 1978, *The Language of Reason*, Pergamon Press, Rushcutters Bay.

Samson D. C. 1993, *Editing Technical Writing*, Oxford University Press, New York.

Selzer J. 1983, "What Constitutes a 'Readable' Technical Style?" in P. V. Anderson, R. J. Brockmann & C. R. Miller (eds), *New Essays In Scientific and Technical Communication: Research, Theory and Practice*, Baywood, New York, pp. 71–89.

Stark, H. A. 1988, "What do Paragraph Markings do?", *Discourse Processes*, vol. 11.

*Style Manual for Authors, Editors and Printers* 2002, John Wiley & Son, Canberra, 6th edn.

Tebeaux E. 1997, *The Emergence of a Tradition: Technical Writing in the English Renaissance, 1475–1640*, Baywood, Amityville.

Trudgill P. 1999, *The Dialects of England*, Blackwell, Oxford, 2nd edn.

van de Velde C. & von Grünau M. 2003, "Tracking Eye Movements while Reading: Printing Press versus the Cathode Ray Tube", *Perception*, 2003, http://www.perceptionweb.com/abstract.cgi?id=v031179. Viewed 15 January 2008.

Wheildon C. 2005, *Type & Layout: Are You Communicating or Just Making Pretty Shapes?*, Worley, Mentone.

Woods B., Moscardo G. & Greenwood T. 1998, "A Critical Review of Readability and Comprehensibility Tests", *The Journal of Tourism Studies*, pp. 49–61.

Zimmerman D. & Clark, D. 1987, *The Random House Guide to Technical and Scientific Communication*, Random House, New York.

# Index